高·等·学·校·教·材

轻化工程
专业英语

郑春玲　主　编
袁爱琳　张正宇　孙　戒　副主编

化学工业出版社
·北京·

内容提要

《轻化工程专业英语》共分五部分，第一部分系统介绍了轻化工程专业涵盖的有机化合物的英文命名体系和方法；第二部分强调了轻化工程领域需重点关注的有机化学、高分子化学、表面化学等基础化学学科的专业英语知识；第三部分为轻化工程纺织化学与染整、助剂与添加剂、皮革、造纸等相关研究方向的科技阅读材料；第四部分为与轻化工程专业相关的应用性英语表达；附录为轻化工程专业相应领域的英文表达规律。内容编排上能够让学生全面、系统地学习到基础而又符合时代发展要求的轻化工程专业英语知识，为查阅专业文献、发表学术论文和加强专业科技交流打下基础。

《轻化工程专业英语》可作为轻化工程及相关专业学生的专业英语教材，也可供相关专业的科技人员及中等英语水平的读者自修参考。

图书在版编目（CIP）数据

轻化工程专业英语/郑春玲主编. —北京：化学工业出版社，2020.9
ISBN 978-7-122-37250-5

Ⅰ.①轻… Ⅱ.①郑… Ⅲ.①轻工业-化工工程-英语-高等学校-教材 Ⅳ.①TQ02

中国版本图书馆CIP数据核字（2020）第103211号

责任编辑：刘志茹　崔俊芳　　　　　　　　装帧设计：关　飞
责任校对：刘　颖

出版发行：化学工业出版社（北京市东城区青年湖南街13号　邮政编码100011）
印　　装：三河市双峰印刷装订有限公司
787mm×1092mm　1/16　印张9　字数219千字　2020年10月北京第1版第1次印刷

购书咨询：010-64518888　　　　　　　售后服务：010-64518899
网　　址：http://www.cip.com.cn
凡购买本书，如有缺损质量问题，本社销售中心负责调换。

定　价：28.00元　　　　　　　　　　　　　　　　　　　版权所有　违者必究

前　言

"轻化工程"是1998年教育部本科专业目录修订后，国内首次出现的高等学校工学类本科专业名称。轻化工程学科对纤维类材料，特别是对天然动植物资源中的纤维类材料的化学成分、组织结构、性能应用等要素进行深入研究，由纺织化学与染整工程、添加剂化学与工程、皮革工程、制浆造纸工程四个专业方向组成。

本教材编写的主要特点如下。

（1）一般的专业英语教材比较偏重专业资料阅读能力训练，而本教材除提供专业阅读资料外，更加注重听、说、译的训练。为此，教材中提供了不少可帮助读者听、说专业英语口语的材料。如在附录中给出了"常用数学符号、数学式、化学式及希腊字母的英文读法"等，这些将为读者进行专业科技交流提供基础。同时也使本教材可作为双语课程、在线课程教材选用。

（2）一般的专业英语教材比较侧重科技英语的翻译，这部分内容学生在大学英语课程学习中已基本涉及，因而本教材对学术英语论文的写作方法进行了一定的补充修改，并将这部分列入附录，供学生在毕业论文环节中作为英文论文或摘要写作的参考模板。

（3）本教材编写过程中，注重从过程工程、大化工的角度选择专业阅读材料，从而形成了覆盖面广、交叉度高的专业英语学习体系，便于读者扩大知识面和掌握更多专业词汇。同时，本教材提供了相关领域的常用词汇表，书末总结了"常见聚合物的中英文名称及缩写""常见玻璃仪器的中英文名称对照"等附录供读者使用，也可作为课内讲座或课外资料选用。

（4）从强化学生自主学习、任课老师更好考核学习效果的角度考虑，本教材中的每篇课文后都列有生词表，有利于学生预习、复习及任课老师的综合考查。

本教材内容包括五部分，第一部分系统介绍轻化工程专业涵盖的有机化合物的英文命名体系和方法，巩固总结专业词汇的构词特点及规律；第二部分为轻化工程专业涉及的化学基础知识，尤其突出有机化学、高分子化学、表面化学等，作为学生从基础英语过渡到专业英语的桥梁；第三部分为轻化工程各专业方向的科技阅读材料，系统介绍与轻化工程纺织化学与染整、助剂与添加剂、皮革、造纸等相关研究方向的基本理论、技术工艺、生产设备、性能检测等内容；第四部分为与轻化工程专业相关的应用性英语表达，如外贸实务、会议投稿等，为读者的就业深造等提供相应的英语训练；附录为轻化工程专业相应领域的常见英文表达规律。

本教材由南京工业大学郑春玲、袁爱琳、张正宇、孙戒、万嵘、王国伟、王海英、迟波，武汉纺织大学姜会钰等老师共同编写。郑春玲任主编，袁爱琳、张正宇、孙戒任副主编。全书由郑春玲统稿、审核。

本书能够顺利完成编写与出版工作，离不开各位领导和同仁的关心、帮助和支持。感谢南京工业大学教务处管国锋处长、陈新民处长的支持；感谢硕士研究生刘明、陈灿、徐彬、厉巧萍以及轻化工程多位本科生为本书编辑资料、录入文字等做出的大量工作。作者在此由衷地向上述人员以及其他关心、支持本书出版的所有人士表示感谢。

由于编者水平有限，难免存在不妥与纰漏，恳请读者批评指正并提出宝贵意见。意见可直接联系主编邮箱：174187350@qq.com。

编者

2020 年 03 月

Contents

Part 1 Nomenclatur Procedures ·· 1

    Lesson 1   Alkanes（烷烃）·· 1

    Lesson 2   Alkenes（烯烃）·· 5

    Lesson 3   Alkynes（炔烃）·· 6

    Lesson 4   Aromatic Hydrocarbons，Arenes（芳香烃）·············· 7

    Lesson 5   Halogeneted Hytrocarbons（卤代烃）······················ 7

    Lesson 6   Compounds Containing Oxygen（含氧化合物）·········· 8

    Lesson 7   Compounds Containing Nitrogen（含氮化合物）········ 15

    Lesson 8   Sulfur Compounds（含硫化合物）························· 18

    Lesson 9   Phosphorus Compounds（含磷化合物）·················· 21

Part 2 General Chemistry in Light Chemical Engineering ············ 24

    Lesson 1   Organic Chemistry ··· 24

    Lesson 2   Polymer Chemistry ··· 27

    Lesson 3   Surface and Interface Chemistry ······················· 37

Part 3 Professional Literature ·· 45

    Unit 1 Dyeing and Finishing ·· 45

      Lesson 1   Dyes ·· 45

      Lesson 2   Fibers ·· 48

      Lesson 3   Color ·· 57

      Lesson 4   Dyeing Principles ·· 62

      Lesson 5   Preparation for Dyeing and Finishing ··············· 68

      Lesson 6   Dyeing and Printing ······································· 80

      Lesson 7   Finishing ··· 89

    Unit 2 Auxiliary and Additive ··· 91

      Lesson 8   Food Additives ··· 91

      Lesson 9   Auxiliaries in Makeup Products ······················· 95

    Unit 3 Leather Manufacture ·· 99

  Lesson 10 Leather—Extraordinary Product of Nature ········· 99
  Lesson 11 Leather Technology ········· 101
 Unit 4 **Pulp and Paper Engineering** ········· 104
  Lesson 12 Pulp Process and Pulp End Uses ········· 104
  Lesson 13 Papermaking ········· 105

Part 4 Applied Literature ········· 107

 Lesson 1 Application Letter ········· 107
 Lesson 2 Recommendation Letter ········· 107
 Lesson 3 Resume ········· 108
 Lesson 4 Notification ········· 109
 Lesson 5 Letter of Inquiry ········· 109
 Lesson 6 Invitation Letter ········· 110
 Lesson 7 Meeting Reply ········· 110
 Lesson 8 Thank-you Note after Meeting ········· 111
 Lesson 9 Inquiry ········· 111
 Lesson 10 Offer ········· 112

Appendix ········· 113

 Appendix 1 Grammar and Translation Features of English for Science and Technology ········· 113
 Appendix 2 English Word Formation Rules of Common Chemical Names ········· 125
 Appendix 3 Names and Abbreviations of Common Polymers ········· 129
 Appendix 4 Speaking of Common Molecular Formulas, Mathematical Symbols and Greek Alphabet ········· 133
 Appendix 5 Common Glassware Names ········· 135

Reference ········· 137

Part 1
Nomenclatur Procedures

Lesson 1　Alkanes（烷烃）

1.1　Straight Chain Alkanes（直链烷烃）

烃的英文构词方法为：hydrogen（氢）取词头，连接 carbon（碳），如
$$\text{hydro-接-carbon} = \text{hydrocarbon}$$
它是可数名词，复数需加"s"做词尾。英文是粘连构词，中译"烃"则是拼读。如烷烃，alkanes；饱和脂肪烃，saturated aliphatic hydrocarbons。

1.1.1　数词

与英语常用数词全然不同，全部化学数词在命名中只用作前缀。应单做记忆的数目英汉对照见表 1-1。

这些词头在化合物命名中的使用规则及本身构词的顺序和英汉对译中，应注意以下几点。

① 由一到四的词头，在作为主链母体词头时，英汉对照都只许选用：

英文名　　　　metha　etha　propa　buta
中译名　　　　甲　　　乙　　丙　　　丁

而作侧链基团数目词头则只准用：

英文名　　　　mono　di（bi）　tri　tetra
中译名　　　　一　　二　　　三　　四

② 由五至十，不论主链碳数或侧链基团数，其英文词头都用由 penta 到 deca，但中译时，主链用"天干"词序戊、己……癸；侧链基团中译仍用数词五、六、七……十，不得混淆。

例如：结构为：

主链词头：buta（丁）；
侧链词头：di（bi，二）

表1-1 英文化学名词中常用数目词头

数目	英文词头	中译词头	数目	英文词头	中译词头
1/2	hemi,semi	半	22	docosa	二十二
1	metha,mono hen,uni	甲,单一个	23	tricosa	二十三
1,1/2	sesqui	倍半	24	tetracosa	二十四
2	etha,di,bi bis,do	乙,二,双,两个	25	pentacosa	二十五
3	propa,tri,tris	丙,三,三个	26	hexacosa	二十六
4	buta,tetra,quadri	丁,四	27	heptacosa	二十七
5	penta,quinque,quinqui	戊,五	28	octacosa	二十八
6	hexa,sexi	己,六	29	nonacosa	二十九
7	hepta,septi	庚,七	30	triaconta	三十
8	octa	辛,八	31	hentriaconta	三十一
9	nona,ennea	壬,九	40	tetraconta	四十
10	deca	癸,十	42	dotetraconta	四十二
11	hendeca,undeca	十一	50	pentaconta	五十
12	dodeca	十二	53	tripentaconta	五十三
13	trideca	十三	60	hexaconta	六十
14	tetradeca	十四	64	tetrahexaconta	六十四
15	pentadeca	十五	70	heptaconta	七十
16	hexadeca	十六	75	pentaheptaconta	七十五
17	heptadeca	十七	80	octaconta	八十
18	octadeca	十八	86	hexaoctaconta	八十六
19	nonadeca	十九	90	nonaconta	九十
20	eicosa	二十	97	heptanonaconta	九十七
21	heneicosa	二十一	100	hecta	一百

③ 英文数目词头从10开始用"-deca"表示"十"的基数。从11到19加前缀，构词为：

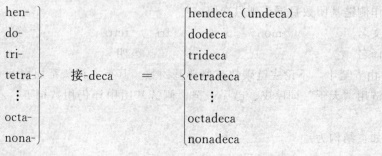

④ eicosa 是二十的词头，heneicosa 二十一；但从 22 至 29，则以 "-cosa" 表示 "二十" 的基数。规律同③，如：

$$\left.\begin{array}{l}\text{do-}\\ \text{tri-}\\ \vdots\\ \text{nona-}\end{array}\right\} \text{接-cosa} = \left\{\begin{array}{l}\text{docosa}\\ \text{tricosa}\\ \vdots\\ \text{nonacosa}\end{array}\right.$$

⑤ 从 30 起，基数 "十" 用 "-aconta" 为代号，前缀数词。构词公式如下：

$$\left.\begin{array}{l}\text{tri-}\\ \text{tetra-}\\ \vdots\\ \text{octa-}\\ \text{nona-}\\ \text{ennea-}\end{array}\right\} \text{接-（a）conta} = \left\{\begin{array}{ll}\text{triaconta} & \text{三十}\\ \text{tetraconta} & \text{四十}\\ \vdots & \text{几十}\\ \text{octaconta} & \text{八十}\\ \left.\begin{array}{l}\text{nonaconta}\\ \text{enneaconta}\end{array}\right\} & \text{九十}\end{array}\right.$$

⑥ 若 30 以后又有个位数，则将个位数词头加十位数前缀之。
例如：tritriaconta 三十三；pentahexaconta 六十五。

1.1.2 直链烷烃构词

直链烷烃的构词法是：按直链烷烃的碳原子总数选择相应的数目词作词头，去掉词尾 "-a"，接粘上烷后规定词尾 "-ane"，即：

$$\text{词头数词去尾 "-a" 接 "-ane"} = \text{某烷}$$

例如：

$$\left.\begin{array}{l}\text{meth（a）}\\ \text{but（a）}\\ \text{dec（a）}\\ \text{hendec（a）}\\ \text{tricos（a）}\\ \text{dotriacont（a）}\\ \text{hexatetracont（a）}\end{array}\right\} \text{接-ane} = \left\{\begin{array}{ll}\text{methane} & \text{甲烷}\\ \text{butane} & \text{丁烷}\\ \text{decane} & \text{癸烷}\\ \text{hendecane} & \text{十一烷}\\ \text{tricosane} & \text{二十三烷}\\ \text{dotriacontane} & \text{三十二烷}\\ \text{hexatetracontane} & \text{四十六烷}\end{array}\right.$$

1.2 Branched Chain Alkanes（支链烷烃）

烷基为简单直链时，按碳原子数选出对应数词词头，去掉词尾 "-a" 换成 "-yl"。公式：

$$\text{数词词头去尾 "-a" 换 "-yl"} = \text{某烷基}$$

例如：

$$\left.\begin{array}{l}\text{meth（a）}\\ \text{eth（a）}\\ \vdots\\ \text{pent（a）}\end{array}\right\} \text{接换 "-yl"} = \left\{\begin{array}{ll}\text{methyl} & \text{甲基}\\ \text{ethyl} & \text{乙基}\\ \vdots & \text{某基}\\ \text{pentyl} & \text{戊基}\end{array}\right.$$

直链烷基构词法常用两种前缀。以下各例不带※号的是 IUPAC 命名。

结构	英文名	中译名
$\overset{4}{C}H_3-\overset{3}{C}H_2-\overset{2}{C}H_2-\overset{1}{C}H-$ 　　　　　　　　　　$|$ 　　　　　　　　　CH_3	1-methylbutyl	1-甲基丁基
$CH_3-CH_2-CH_2-\overset{*}{C}H-$ 　　　　　　　　　　$|$ 　　　　　　　　　CH_3	sec-pentyl	仲戊基
$\overset{4}{C}H_3-\overset{3}{C}H-\overset{2}{C}H_2-\overset{1}{C}H_2-$ 　　　　$|$ 　　　CH_3	3-methybutyl	3-甲基丁基
$CH_3-\overset{*}{C}H-CH_2-CH_2-$ 　　　　$|$ 　　　CH_3	iso-pentyl	异戊基
$\overset{3}{C}H_3-\overset{2}{C}H_2-\overset{1}{C}-$ 　　　　　　　$|$ 　　　　　　CH_3 (CH_3 above)	1,1-dimethypropyl	1,1-二甲基丙基
$CH_3-CH_2-\overset{*}{C}-$ 　　　　　$|$ 　　　　CH_3 (CH_3 above)	tert-pentyl	叔戊基
$\overset{3}{C}H_3-\overset{2}{C}-\overset{1}{C}H_2-$ 　　$|$ 　CH_3 (CH_3 above)	2,2-dimethylpropyl	2,2-二甲基丙基
$CH_3-\overset{*}{C}-CH_2-$ 　　$|$ 　CH_3 (CH_3 above)	neo-pentyl	新戊基

注意：※结构是按烷基的总碳原子数命名。※处的碳原子为 2 就用"sec-"前缀形容，中译为"仲"或"另"；为 3 时用"tert-"作前缀，中译为"叔"；为 4 时用"neo-"，中译为"新"；结构为 $CH_3-CH-\cdots\cdots$ 统称前缀"iso-"，中译"异"。这些都是常用但不正规的
　　　　　　　　　　　　　　　$|$
　　　　　　　　　　　　　CH_3
命名。严格地说，未加※号的英汉命名才是 IUPAC 系统命名，特别是基团复杂时用最好，因为准确定位。但化合物结构较简单时，多采用前者或混用。

1.3　Cycloalkanes Alicyclic hydrocarbons（环烷烃）

1.3.1　简单环烷烃

简单环烷 IUPAC 命名英文构词方法为：

$$\text{cyclo-接直链对应烷全词} = \text{cyclo}\cdots\cdots\text{ane}$$

结构与英汉对译比较如下：

　　　　　链烷　　　　　　　　　　　　　　　环烷
　　　$CH_3CH_2CH_3$

propane 丙烷　　　　　　　　　　cyclopropane 环丙烷
CH₃(CH₂)₂CH₃
butane 丁烷　　　　　　　　　　cyclobutane 环丁烷
CH₃(CH₂)₃CH₃
pentane 戊烷　　　　　　　　　　cyclopentane 环戊烷
CH₃(CH₂)₄CH₃
hexane 己烷　　　　　　　　　　cyclohexane 环己烷

例如： 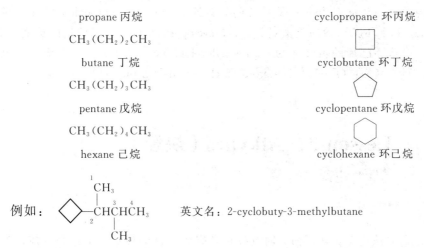 英文名：2-cyclobuty-3-methylbutane

这是按英文基团词头字母顺序编号的。但是中译名按我国规定的先简后繁原则，应该为 2-甲基-3-环丁基丁烷。

1.3.2　桥环（Bridged rings）

简单桥环烃的构词顺序和规定是：词头写总环数，用 bicyclo-、tricyclo- 等表示二环、三环。接写方括号 [ ]。括号中用阿拉伯数字由大到小依次表示各桥上的桥墩碳原子数，但不包括桥头碳原子数在内。最后接总环中全部碳原子数所对应的直链烷作词尾。顺序编号由桥头开始沿长桥绕中桥至短桥结束。如：

英文名：1,2-dimethyl-5-ethylbicyclo [2.1.0] pentane
中译名：1,2-二甲基-5-乙基二环 [2.1.0] 戊烷

英文名：spiro [3.4] octane
中译名：螺 [3.4] 辛烷

Lesson 2　Alkenes（烯烃）

烯烃的 IUPAC 系统命名规则如下。
① 选择结构中含双键（double bond）的最长碳链作为命名母体。
② 母体把相应烷的词尾 "-ane" 换成 "烯" 的词尾 "-ene"（普通命名则为 -ylene）。
③ 给母体链编号以确定双键位置，给它以最小序号。
④ 若有侧链烃基，用前缀指明并标出序号。位置唯一时则序号可省去。

如乙烯的英文名为：ethene（ethylene）。丙烯的英文名为：propene（propylene）。1-丁烯的英文名为：1-butene。1,3-环戊二烯的英文名为：1,3-cyclopentadiene。环辛四烯的英文名为：cyclooctatetraene。2-甲基环己烯的英文名为：2-methylcyclohexene。（E）-1,4-二氯-3-（2-氯乙基）-2-甲基-2-戊烯的英文名为：（E）-1,4-dichloro-3-(2-chloroethyl)-2-methyl-2-pentene。

Lesson 3　Alkynes（炔烃）

按 IUPAC 系统命名法规定，炔烃命名时选择包含碳碳叁键的最长碳链作为母体，其构词只把相应的烷词尾"-ane"换成"-yne"即可，中译为"炔"。侧链烃基同烷、烯命名法，做前缀并编号。如 3-环己烷基丙炔的英文名为：3-cyclohexypropyne。

由上所述可知，词尾-ane、-ene、-yne 分别是 IUPAC 系统命名的烷、烯、炔代号。关于其一价基作为基团时的构词变化是：

烷基：词尾-ane 变成-yl；烯基：词尾-ene 变成-enyl；炔基：词尾-yne 变成-ynyl。

由于仍有场合沿用其他命名，故列表 1-2 对照。

表 1-2　某些炔的系统命名和其他命名

结构	IUPAC 命名	其他命名
$HC\equiv CH$	ethyne 乙炔	acetylene 乙炔
$CH_3C\equiv CH$	propyne 丙炔	methylacetylene 甲基乙炔
$C_2H_5C\equiv CH$	1-butyne 1-丁炔	ethylacelene 乙基乙炔
$CH_3C\equiv CCH_3$	2-butyne 2-丁炔	dimethylacetylene 二甲基乙炔
$CH_3CH_2CH_2C\equiv CH$	1-pentyne 1-戊炔	n-propylacetylene 正丙基乙炔
$CH_3CH_2C\equiv CCH_3$	2-pentyne 2-戊炔	methylethylacetylene 甲基乙基乙炔
$CH_3CHC\equiv CH$ 　　$\mid$ 　　CH_3	3-methyl-1-butyne 3-甲基-1-丁炔	isopropylacetylene 异丙基乙炔
$H_2C=CHC\equiv CH$	1-buten-3-yne 1-丁烯-3-炔	vinylacetylene 乙烯基乙炔
$H_2C=CHC\equiv CCH=CH_2$	1,5-hexadien-3-yne 1,5-己二烯-3-炔	divinylacetylene 二乙烯基乙炔
$C_{13}H_{24}$	tridecyne 十三(碳)炔	

Lesson 4　Aromatic Hydrocarbons，Arenes（芳香烃）

（1）硝基、烷基、卤素等简单一元取代，以取代基作词头，母体用烃全词。

如硝基苯的英文名为：nitrobenzene。（1,1-二甲基乙基）苯（叔丁基苯）的英文名为：(1,1-dimethylethyl) benzene (tert-butylbenzene)。甲苯的英文名为：toluene (methylbenzene)。苯酚的英文名为：phenol (benzenol)。苯乙烯（乙烯苯）的英文名为：styrene (ethenylbenzene, phenylethene)。苯胺的英文名为：aniline (benzenamine)。

（2）按照 IUPAC 命名法则，硝基、烷基、卤素等小基团的简单多元取代，一般多以芳烃全词作母体放在后面，前面用编号前缀基团，并把小序号让给主体取代基；或编号由前缀"ortho-""meta-""para-"分别表示邻、间、对位，缩写用小写 o-、m-、p- 表示。

如邻二氯苯的英文名为：o-dichlorobenzene。对碘硝基苯（4-碘硝基苯）的英文名为：p-iodonitrobenzene。对硝基甲苯（4-硝基甲苯）的英文名为：p-nitrotoluene。

（3）按照 IUPAC 命名法则，芳烃的基团复杂取代命名如间氯苯酚（3-氯苯酚）的英文名为：m-chlorophenol (3-chlorophenol)。2-溴苯甲醚（2-溴-1-甲氧基苯）的英文名为：2-bromo-1-methoxybenzene。3-溴苯磺酸（间溴苯磺酸）的英文名为：3-bromophenylsulfonic acid (m-bromobenzene sulfonic acid)。

有几个地方请注意：

非芳香烃变烃基的构词一般是烷去词尾"-ane"，烯、炔去"-e"换成"-yl"。但在芳烃中，苯的对应基是 phenyl，而 benzenyl 是 $C_6H_5C\equiv$（苯亚甲基，次苄基）；benzyl 是 $C_6H_5CH_2-$（苯甲基或苄基）。这些又极易与 benzoyl，C_6H_5CO- 苯甲酰基相混淆。另外稠环芳烃有取代基时，必须按表中既定编号处理，不得任意编排。如9-溴蒽的英文名称只能写成：9-bromoanthracene。

Lesson 5　Halogeneted Hytrocarbons（卤代烃）

卤族元素名称分别是：氟（fluorine）、氯（chlorine）、溴（bromine）、碘（iodine）。

若在化合物中，英文构词做词头时，构词方法为：

元素名词去"-ine"换成"-o"

即 fluoro-、chloro-、bromo-、iodo-为氟、氯、溴、碘英文词头。

卤素做词尾时，构词方法为：

元素名词去"-ine"换成"-ide"

即 fluoride、chloride、bromide、iodide 为氟、氯、溴、碘（化物）的英文词尾。

既然卤代烃由烃基与卤素两部分组成，且卤素既可作词头也可做词尾，相应的烃基就同样既可作词尾也可作词头了。

(1) 卤代烃的两套命名法

① IUPAC 命名构词法：

<u>数词接卤素词头接烃全词 ＝ 某卤代烃</u>

必要时编号，原则与烃一致。值得注意的是，卤素的英文词序是按照 Br、Cl、F、I 的字母顺序，但中译时应按我国规定为氟、氯、溴、碘，并非完全字字相扣。

② 其他命名的构词法有：

<u>烃基词头＋数词接卤素词尾 ＝ 某基卤（化物）</u>

(2) 命名实例对照

一氯甲烷（甲基氯）的 IUPAC 名称为：chloromethane，俗名：methyl chloride。四氯甲烷（四氯化碳）的 IUPAC 名称为：tetrachloromethane，俗名：carbon tetrachloride。

而 2,4-二硝基氟苯的 IUPAC 名称为：2,4-dinitrofluorobenzene，俗名：2,4-dinitrophenyl fluoride。聚四氟乙烯（特氟龙）的 IUPAC 名称为：polytetrafluoroethene，俗名：polytetrafluoroethylene（Teflon）。

Lesson 6　Compounds Containing Oxygen（含氧化合物）

1.6.1　Alcohols，Phenols（醇，酚）

烃上连接羟基—OH ［hydroxy(l)］一般就命名为醇或酚。从结构上识别，羟基在芳环上中译名为酚，在非芳环上则中译名为醇。现分述其命名构词如下。

(1) 醇的命名构词

① IUPAC 构词法

$$烃词\begin{cases}-ane\\-ene\\-yne\end{cases} 去尾 "-e" 接 "-ol" 词尾 ＝ 某醇$$

这是一元醇公式，必要时可标序号。例如甲醇的英文名为：methanol。环己醇的英文名为：cyclohexanol。

若系多元醇，则烃不去词尾 e 而用全词，并在 -ol 之前加数目词头。即：

<u>烃全词接数词头再接 "-ol" ＝ 某几醇</u>

如乙二醇的英文名为：1,2-ethanediol。环己六醇的英文名为：cyclohexanehexaol。

在复杂的情况下，处理如下：

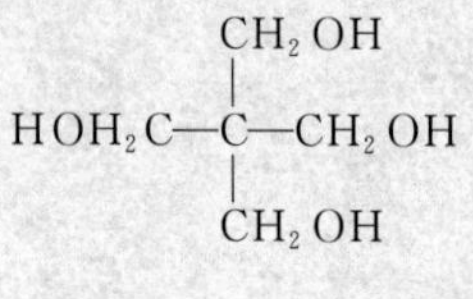

$$\begin{array}{c}CH_2OH\\|\\HOH_2C-C-CH_2OH\\|\\CH_2OH\end{array}$$

英文名：2,2-bis(hydroxymethyl)-1,3-propanediol

中译名：2,2-双(羟甲基)-1,3-丙二醇

② 其他命名

有章可循的一元醇构词法为：

<u>烃基词头＋alcohol ＝ 某一元醇</u>

如甲醇的英文名为：methyl alcohol。异丙醇的英文名为：isopropyl alcohol。而有些多元醇不仅可采用 IUPAC 命名法命名，还具有俗名。例如乙二醇（甘醇）的 IUPAC 名称为：ethanediol，俗名：glycol。丙三醇（甘油）的 IUPAC 名称为：propanetriol，俗名：glycerol。

(2) 酚的命名构词

酚类最有规律的 IUPAC 构词法为：

<u>芳香基去词尾-yl 变为-ol ＝ 某酚</u>

例如：苯基是 phenyl，去-yl 换成-ol 即成 phenol，这就是苯酚。萘酚等亦照此办理：naphthol。不过要注意的是，在《Chemical Abstracts》中仍用 benzenol 作苯酚，构词类同于醇。

多元酚的 IUPAC 构词类同于多元醇：

<u>芳烃全词接数词词头再接-ol＝ 多元酚</u>

例如：benzenediol 代表苯二酚。当然，在指明其具体结构时，还需在前面注明序号，或用 o-、m-、p-标出。

多元酚的普通命名规则是把它看成烃的衍生物，即芳烃作母体，如：

<u>数目词头接 hydroxy 接芳烃全词 ＝ 多元酚</u>

例如：间苯二酚的普通命名是：

m-dihydroxybenzene 或 1,3-dihydroxybenzene，但中译时最好译成间苯二酚或 1,3-苯二酚。千万不要死扣硬译成：间二羟苯或 1,3-二羟基苯。

(3) 醇盐与酚盐（Alkoxides）

醇、酚都能与活泼金属或碱土金属发生置换反应而生成结构类似 ROM、$(RO)_2M$ 的化合物。式中 M 代表一、二价的金属元素；RO—代表烃氧基。所以，实际上是可将该烃氧基化物看成弱酸强碱盐，其命名可有两种：

<u>金属元素全词＋alkoxide ＝ 醇盐或酚盐</u>

或把醇、酚-ol 的变成-olate，即：

<u>金属元素＋醇、酚全词接-ate ＝ 酚或醇盐</u>

例如乙醇钠的英文名为：sodium ethoxide，或 sodium ethanolate。苯酚钾的英文名为：potassium phenoxide，或 potassium phenolate。

1.6.2　Ethers（醚）

(1) 普通命名

我国关于醚的命名原则是：以"醚"字作母体词尾，前面把两个烃基写上。当两个基团相同时，命名为"二某基醚"或省去"二"字。当两个基团不同时，则按先小后大的原则排列。我国这套构词跟英文普通命名基本一致，所不同的是，英文烃基按词头字母排列之先后顺序排列。命名方法如下。

当二烃基一样时，英文普通命名：

<u>di 烃基＋ether ＝ （二）某醚</u>

当二烃基不一样时，英文普通命名：

先词头烃基＋后词头烃基＋ether ＝ 某醚

如二乙醚的英文名为：diethyl ether。甲乙醚的英文名为：ethyl methyl ether。甲基正丙基醚的英文名为：methyl *n*-propyl ether。

(2) IUPAC 命名

本命名原则的显著特点是，特征词"ether"已不复存在，而强调选择最长碳链或复杂碳链作为贯穿全局的母体。具体办法如下：

① 在 R—O—R′中，比较两个基团的碳链，看哪一个最长或最复杂，则选其作为母体，并做词尾。也就是说，烷、烯、炔、芳烃（-ane、-ene、-yne、-arene）等都可作词尾，用它们取代了 ether 这个中文对等的"醚"字。

例如：$CH_3OC_2H_5$、$CH_3OC_6H_5$、HO—△—OCH_3 的词尾分别应为：-ethane、-benzene、-cyclopropanol。

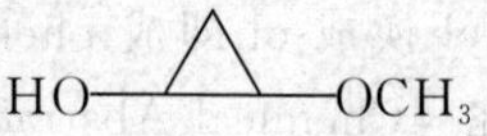

② 剩下的小的，或简单的 RO—，可看成一个烃氧基整体（alkoxy-）用作前缀。烃氧基的构词公式为：

烃基去词尾"-yl"换成"-oxy"＝ 烃氧基

例如烃氧基的英汉对照为：

$CH_3O—$	methoxy-	甲氧基
$CH_2=CH—O—$	ethenoxy-	乙烯氧基
$C_6H_5O—$	phenoxy-	苯氧基

所谓杂原子（hetero atoms）是指代替骨架链节上碳原子位置的其他原子。用词头 oxa-、thia-、azo-、phospha-表氧杂、硫杂、氮杂、磷杂。词尾根据母体而定，用全词。

如氧环戊烷的 IUPAC 名称为：oxacyclopentane [tetrahydrofuran(e) ※]，普通命名：oxolane（少用）。氧环丙烷（噁丙烷，氧化乙烯）的 IUPAC 名称为：oxacyclopropenane，其他命名：oxirane（ethylene oxide）。氧（杂）环丁烷的 IUPAC 名称为：oxacyclohexane，其他命名：oxetane。氧环己烷（四氢吡喃）的 IUPAC 名称为：oxacyclohexane [tetrahydropyran(e)]。呋喃的 IUPAC 名称为：furan(e)。

冠醚可以看成是乙二醇的缩合物（condensed polymer），是大环多醚（large-ring polyethers）一类。其命名公式为：

x-crown-y ＝ 冠醚

其中，"x"指环上总原子数，"y"代表环节上的氧原子数。如 15-冠醚-5 的英文名为：15-crown-5。

1.6.3 Aldehydes, Ketones（醛，酮）

(1) 醛的命名

① 普通命名

醛的传统构词从相应的酸演变而来，采用如下方法：

酸去-ic (-oic) acid 换成 -aldehyde ＝ 某醛

如甲醛（formaldehyde）根据甲酸（formic acid）演变，乙醛（acetaldehyde）根据醋酸（acetic acid）演变，苯甲醛（benzaldehyde）根据苯甲酸（benzoic acid）演变而来。

② IUPAC 系统构词

醛的 IUPAC 系统构词按规定由相应的烃演变而来，采用如下方法：

烃去词尾-e 变成-al ＝ 相应的一元醛

如乙醛的 IUPAC 名称为：ethanal。3-甲基丁醛的 IUPAC 名称为：3-methylbutanal。丙烯醛的 IUPAC 名称为：propenal。

简单的醛都省去"1"序号。对于复杂醛，可分以下几种情况构词。

其一，链状二醛，采用公式：

烃全词接-dial ＝ 某二醛

如 2-丁炔二醛的英文名为：2-butynedial。

其二，遇复杂情况，—CHO 可以"formyl""甲酰基"作为词头，或用"oxo-"（氧代）做词头。

如 3-甲酰苯甲酸的英文名为：3-formylbenzoic acid。3-氧代丙酸（丙醛酸）的英文名为：3-oxopropanoic acid。

（2）酮的命名

① 普通命名

酮的普通命名以羰基单独作词尾，"ketone"中译为"（甲）酮"。词头按两侧的烃基，烃基按词头字母顺序排列，中译则先小后大。采用方法如下：

烃基＋烃基＋ketone ＝ 某某（甲）酮

如甲乙酮的英文名为：ethyl methyl ketone。甲基苯基酮的英文名为：methyl phenyl ketone。丙酮（二甲酮）的英文名为：acetone (dimethyl ketone)。

② IUPAC 构词规则

酮的 IUPAC 构词，原则上选择包括羰基在内的最长碳链作为命名基础。具体构词方法为：

烃去词尾-e 换成-one ＝ 相应的酮

如 5-乙基-3-庚酮的英文名为：5-ethyl-3-heptanone。4-甲基环戊酮的英文名为：4-methylcyclopentanone。4-戊烯-2-酮的英文名为：4-penten-2-one。

多元酮则在烃全词后插入数词头接酮。如 2,4-戊二酮的英文名为：2,4-pentanedione。

（3）醌的命名

醌（Quinones）实际上是环状共轭的二酮结构，其构词以"-quinone"做词尾，芳烃做词头。如 1,2-苯醌（3,5-环己二烯-1,2-二酮）的英文名为：1,2-benzoquione (3,5-cyclohexadiene-1,2-dione)。9,10-蒽醌的英文名为：9,10-anthraquinone。

（4）酮、醛反应的产物

① 醛或带甲基的酮与 HCN 的加成产物有两种命名法。普通构词法为：

an aldehyde ＋ hydrocyanic ⟶ a cyanohydrin
醛或酮全称＋cyanohydrin ＝ 醛或酮氰醇

IUPAC 构词以—CN 在内的最长碳链为母体，按"腈"来命名的。构词法为：

烃全词接-nitrile ＝ 某腈

羰基作为前缀，当取代基。如：

HCHO＋HCN ⟶ CH$_2$—CN
｜
OH

产物的 IUPAC 命名为：2-hydroxyethanenitrile（2-羟基乙腈），普通命名为：formadehyde cyanohydrin（甲醛氰醇）。

② 与亚硫酸氢钠（sodium bisulfite）反应，生成的产物结构是磺酸盐。构词如下：

$$\text{金属全词} + \text{酸盐词尾} = \text{某酸盐}$$

如：

$$\begin{array}{c} CH_3CH_2 \\ \diagdown \\ C=O \\ \diagup \\ H_3C \end{array} + NaHSO_3 \longrightarrow \begin{array}{c} CH_3CH_2 \quad OH \\ \diagdown \diagup \\ C \\ \diagup \diagdown \\ H_3C \quad SO_3^-Na^+ \end{array}$$

产物的 IUPAC 命名是：sodium 2-hydroxy-2-butylsulfonate（2-羟基-2-丁基磺酸钠）。

③ 与羟胺反应（hydroxyamine），生成产物叫"肟"。构词方法：

$$\text{醛和酮全词} + \text{oxime} = \text{某醛或酮肟}$$

如：

$$\begin{array}{c} CH_3 \\ \diagdown \\ C=O \\ \diagup \\ CH_3 \end{array} + NH_2OH \longrightarrow \begin{array}{c} CH_3 \\ \diagdown \\ C=NOH \\ \diagup \\ CH_3 \end{array}$$

产物的 IUPAC 命名是：propanone oxime（丙酮肟）。

当分子中有优先主官能团时，以"肟基"=NOH 反过来作词头，直译为 hydroxyimino-（羟亚氨基）作为前缀。

④ 与 R—NH$_2$ 反应生成席夫碱（Schiff base），可按胺类或烃命名。如：

$$\text{C}_6\text{H}_{10}=O + H_2N-C(CH_3)_3 \longrightarrow \text{C}_6\text{H}_{10}=N-C(CH_3)_3$$

产物命名是：N-cyclohexylene-2-methyl-2-propanamine（t-butyliiminocyclohexane）（N-亚环己基-2-甲基-2-丙胺[叔丁亚氨基环己烷（混合构词）]。

⑤ 与肼（H$_2$NNH$_2$，hydrazine）、苯肼（H$_2$NNH-C$_6$H$_5$，phenylhydrazine）反应，生成物称作腙（hydrazone）。构词用公式：

$$\text{醛和酮全词} + \text{hydrazone} = \text{某醛或酮腙}$$

如：

环戊酮 + NH$_2$NH$_2$ ⟶ 环戊酮腙

产物命名是：cyclopentanone hydrazine（环戊酮腙）。

$$CH_3CHO + H_2NNHC_6H_5 \longrightarrow CH_3CH=NNHC_6H_5$$

产物命名是：ethanal phenylhydrazone（乙醛苯腙）。

吖嗪"azine"专用于两端相同的酮或醛的二腙形态。如：

$$2\,\text{环戊酮} + NH_2NH_2 \longrightarrow \text{环戊基}=N-N=\text{环戊基}$$

产物命名是：cyclopentanone azine（环戊酮吖嗪）。

与氨基脲（H$_2$NNHCONH$_2$，aminourea，semicarbazide）反应，产物命名为缩氨基脲（半卡巴腙）"semicarbazone"，前留空格，冠以醛或酮全称。例如：

$$C_6H_5CHO + H_2NNHCONH_2 \longrightarrow C_6H_5CH=NNHCONH_2$$

产物命名是：benzaldehyde semicarbazone 苯甲醛缩氨基脲（苯甲醛半卡巴腙）。

1.6.4　Carboxylic Acids（羧酸）及其衍生物

（1）羧酸

① 简单非环羧酸

羧酸的命名，至今国外多用俗名，无书写规律可循。但 IUPAC 命名很严谨：

$$\text{链烃去词尾 -e 换成 -oic} = \text{某酸}$$

如甲酸（蚁酸）的 IUPAC 名称为：methanoic acid，俗名：formic acid。乙酸（醋酸）的 IUPAC 名称为：ethanoic acid，俗名：acetic acid。氯乙酸的 IUPAC 名称为：chloroethanoic acid，俗名：chloroacetic acid。

② 二元羧酸

直链二元羧酸的 IUPAC 构词方法为：

$$\text{烃全词 + di 接 oic acid} = \text{某二元酸}$$

如（Z）-丁烯二酸的 IUPAC 名称为：（Z）-butenedioic acid。

③ 多元羧酸

看实例以识别构词规则：把三个羧基除去就是丙烷骨架。所以：

CH$_2$—COOH
|
CH—COOH　　IUPAC：propane-1,2,3-tricarboxylic acid
|　　　　　　中译名：丙烷-1,2,3-三羧酸
CH$_2$—COOH

三个羧基要等同对待，故不能叫做 3-羧基戊二酸，英文命名也不能写成 3-carboxypentanedoic acid。

对于三个羧基不等同的化合物：

CH$_2$—COOH
|
CH—CH$_2$—COOH
|
CH$_2$—COOH

其 IUPAC 命名应为：2-carboxymethylpentanedioic acid，中译名为：2-羧甲基戊二酸。

④ 挂环羧酸

当羧基（—COOH）直接连在环上时，化合物的 IUPAC 构词法为：

$$\text{环烃全词 + 数词词头接 carboxylic acid} = \text{某羧酸}$$

如环戊烷二甲酸的英文名为：cyclopentane-1,2-dicarboxylic acid。1,2-苯二甲酸的英文名为 benzene-1,2-dicarboxylic acid。

若羧基（—COOH）不直接挂环，则为：

$$\text{环烃基接羧酸全称} = \text{某基某酸}$$

如苯基乙酸的英文名为：phenylethanoic acid。

⑤ 特殊情况

当有复杂主官能团可充当母体时，羧基可看做取代基，以 carboxy(l)- 的形式作为前缀。

如 3-羧甲基苯磺酸钠的名称为：sodium 3-carboxymethylbenzenesulfonate。

（2）酸酐（Anhydrides）

酸酐的构词方法为：

<u>换羧酸词尾 "acid" 为 "anhydride"</u> ＝ 某酸酐

如3-甲基丁酸酐的英文名：3-methylbutanoic anhydride。甲酸乙酸酐的英文名：ethanoic methanoic anhydride。

（3）酰卤（Acyl halides）

酰基名称是以特征基团酰基直链卤素为命名根据的。即：

$$-\overset{O}{\overset{\|}{C}}-X \quad (X 代表卤素)$$

① 酰基构词方法为：

<u>含氧酸去 "-ic acid" 换成 "-yl"</u> ＝ 某酰

所谓酰基，即含氧酸去羟基后的基团，不论是有机含氧酸或无机含氧酸，皆如此构词。请注意以下命名实例各自的对应变化。

<u>formic acid</u> / methanoic acid HCOOH 甲酸

<u>acetic acid</u> / ethanoic acid CH_3COOH 乙酸

<u>sulfuric acid</u> $(HO)_2SO_2$ 硫酸

<u>carbonic acid</u> $(HO)_2CO$ 碳酸

<u>carboxylic acid</u> —COOH 羧酸

formyl / methanoyl HCO— 甲酰

acetyl / ethanoly $CH_3CO—$ 乙酰（<u>aceto</u>）

sulfuryl $\diagdown SO_2 \diagup$ 硫酰

$\diagdown CO \diagup$

carbonyl 碳酰，羧基
carbonyl 甲酰，碳酰

请注意，加下划线的酸和酰是俗名，但已通用；而IUPAC命名（未加下划线）虽然极有规律，但使用得较少，甚至现行的《英汉化学化工词汇》里都查不到很多酰的系统命名。

② 酰卤命名，构词方法为：

<u>酰基＋卤素词尾</u> ＝ 某酰卤

如碳酰氯英文名为：carbonyl chloride。苯甲酰溴英文名为：benzoyl bromide（benzene carbonyl bromide）。

当有复杂主官能团能充当母体命名时，酰卤一词在中文里要变成"卤酰基"（haloacyl-），反作词头前缀。如：

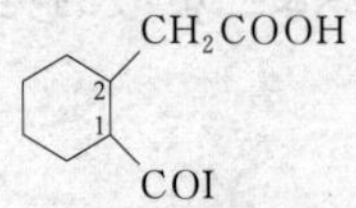

（结构式：环己基，2位连 CH_2COOH，1位连 COI）

英文名：2-iodocarbonylcyclohexylethanoic acid

中译名：2-碘碳酰环己基乙酸

（4）酰胺（Amides）

特征基团为—$CONH_2$ 的羧酸衍生物（derivatives）称为酰胺，其IUPAC构词方法为：

<u>羧酸去词尾 "-ic acid" 或 "-oic acid" 换 "-amide"</u> ＝ 相应的酰胺

如甲酰胺的英文名为：methanamide（formamide）。2-甲基丙酰胺的英文名为：2-methylpropanamide。

若氨基上的氢被取代，构词做如下修饰：

① 取代基简单，用 $N-$ 标定位置前缀。

如 N-甲基-N-乙基乙酰胺的英文名为：N-ethyl-N-methylethanamide。

② 结构复杂，酰胺亦可做前缀。

情况一：结构为 RCONH—，中译名为酰氨基，只将酰胺全词的-amide 变成-amido 作为前缀。例如：

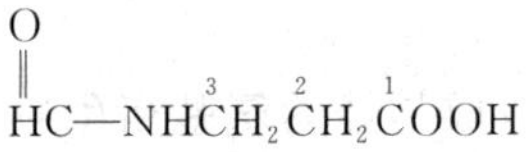

$$\underset{\overset{|}{HC-NHCH_2CH_2COOH}}{\overset{O}{\|}}$$
     3   2   1

英文名：3-methanamidopropanoic acid
中译名：3-甲酰氨基丙酸

情况二：H_2NCO— 作为前缀时，中译名为氨甲酰或氨碳酰基，英文名应为：aminocarbonyl 或 aminoformyl，但 carbamoyl 或 carbamyl 已被通用。例如：

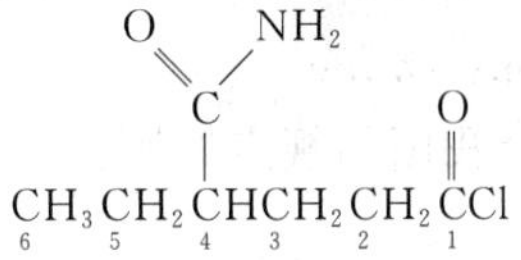

$$CH_3CH_2\overset{NH_2}{\underset{|}{CH}}CH_2CH_2\overset{O}{\underset{\|}{C}}Cl$$
  6   5   4   3   2   1

（式中上方还有 C=O 连到 CH 上）

英文名：4-carbamoylhexanoyl chloride
中译名：4-氨甲酰基己酰氯

（5）酯（Esters）

"esters"是酯的总称，但就具体化合物命名而言，"esters"使用不多。因为酯是醇和酸的缩合产物，所以以酸的衍生物来命名。不论有机或无机含氧酸盐或酯，都用下面这个构词法：

<u>醇烃基（或金属）＋酸词尾 ＝ 酯（盐）</u>
<u>酸词尾：由-ic acid 变成 -ate</u>

如甲酸甲酯的 IUPAC 名称为：methyl methanoate，常用名：methyl formate。乙酸的 IUPAC 名称为：ethyl ethanoate，常用名：ethyl acetate。丙酸丙酯的 IUPAC 名称为：propyl propanoate，常用名：propyl propionate。

Lesson 7　Compounds Containing Nitrogen（含氮化合物）

1.7.1　胺（amines）

在现行文献和书籍中，胺的英汉普通命名和国际系统命名是混用的。因为都有规则，故英文当分别介绍。不过中译时，只用其最常见者，以省篇幅。

（1）基团构词

首先回顾一下无机氨及铵盐的构词：

NH_3：ammonia 氨
NH_4^+：ammonium 铵（阳离子，类金属）

当 NH_3 中的"H"被有机烃基取代时，可能有：RNH_2、R_2NH、R_3N 三种形式的产物。其中，R 可以相同，也可以不相同。这三种产物的中译名词尾母体都叫"胺"，对应的英文名词尾为-amine。做词头时，英文名用 amino-表示氨基，请注意与 amido-表酰氨基的词头加以区别。

出现 R_4H^+ 结构时，命名用 ammonium 铵。

（2）普通命名

以 N 为核心做母体铵，以-amine 做词尾；烃基按照分子量由小到大做词头。构词公式为：

烃基接词尾-amine ＝ 某胺

如乙胺的英文名为：ethylamine。二乙胺的英文名为：diethylamine。甲基乙烯基胺的英文名为：methylethenylamine。

(3) IUPAC 系统命名

方法一：选择最长碳链对应的烃为母体作词尾，从而把胺类看成是烃的氨基衍生物，用 amino-氨基作词头，必要时前缀序号定位。采用如下方法：

（序号）amino-接烃全词 ＝ 某胺

如甲酸甲酯的 IUPAC 名称为：methyl methanoate，常用名：methyl formate。2-丙胺（异丙胺）的 IUPAC 名称为：2-aminopropane，普通命名：isopropylamine。

最好不要直译成 2-氨基丙烷。尽管这是原文意思，但不太符合我国的规定原则。

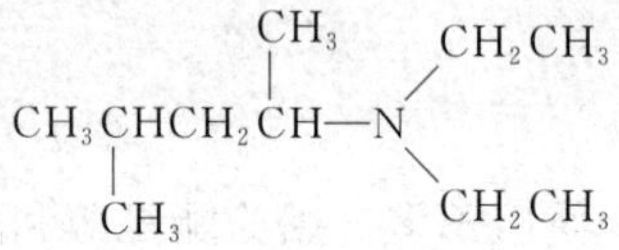

IUPAC：4-(diethylamino)-2-methylpentane
中译名：4-(二乙氨基)-2-甲基戊烷

方法二：选择最长碳链和 N 一起作为母体，母体用烃去掉词尾-e 变成-amine。构词法为：

N-烃基接长链烃-e 变为-amine ＝ N-某基某胺

这样，上例中的结构可命名为：N,N-diethyl-4-methyl-2-pentanamine（N,N-二乙基-4-甲基-2-戊胺）。

(4) 芳香胺的 IUPAC 系统命名

按上述同理，苯胺命名为 benzenamine 是正规的，美国《Chemical Abstracts》至今采用该词。但是，沿用的俗名 aniline 也已得到 IUPAC 的承认，因而后者被更广泛地采用作为命名母体。如 3-溴苯胺的英文名为：3-bromoaniline。N,N-二甲基苯胺的英文名为：N,N-dimethylaniline。

其他芳胺的构词，可按照 IUPAC 脂肪胺办理，普通命名时亦可以芳基作为词头，-amine 作为词尾。如 2-萘胺的 IUPAC 名称为：2-naphthalenamine，普通命名：2-naphthylamine。

同理，芳香胺在有其他复杂基团作为母体主官能团时，则氨基作词头，即用 amino-作为前缀。

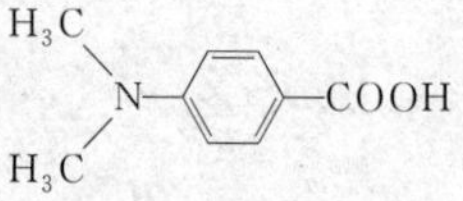

英文名：4-(N,N-dimethylamino)-benzoic acid
中译名：4-(N,N-二甲氨基)苯甲酸

1.7.2 季铵（Quaternary ammoniums）

此类化合物是 $R_4N^+X^-$ 结构状态。实际上是铵上带正电荷（positive charge），X^- 上带负电荷（negative charge）的基团，与无机铵盐相似，所以其命名构词亦相似。采用如下方法：

烃基接-ammonium＋阴离子词尾 ＝ 某铵盐

如氯化铵的英文名为：ammonium chloride。氯化四甲铵的英文名为：tetramethylammonium chloride。

但应注意，伯、仲、叔胺盐的构词最好另用公式，以盐酸盐为例：

胺全词＋hydrochloride ＝ 某胺盐酸盐

例如：

NH_4Cl　　　　　　　　　英文名：methanamine hydrochloride
　　　　　　　　　　　　　中译名：甲胺盐酸盐（盐酸甲胺）
　　　　　　　　　　　　　英文名：hydroxethyltrimethylammonium hydroxide
$[(CH_3)_3N^+CH_2CH_2OH]OH^-$　中译名：氢氧化三甲基羟乙基铵
　　　　　　　　　　　　　俗名：choline（胆碱）

1.7.3　重氮盐与偶氮化合物（Diazonium salts and azo-compounds）

（1）重氮化合物（diazo-compounds）

结构为 Ar—N_2X 或 $-\overset{\overset{N_2}{\|}}{C}-$ 的化合物都叫做重氮化合物。用 diazo-词头表示"重氮"一词，构词接-nium 时，则为其盐类。例如：

芳香伯胺（aromatic primary amines）在低温下与亚硝酸盐的强酸液反应，得到的主产物统称为重氮盐，其结构为 Ar—N_2^+ Y^-，构词采用如下方法：

芳烃全词接 diazomium＋酸根尾 ＝ 某重氮盐

例如：

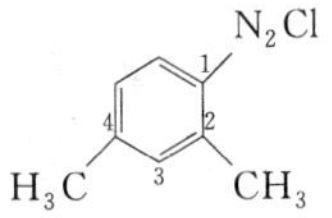

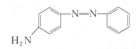
　　　　　英文名：2,4-dimethylbenzenediazonium chloride
　　　　　中译名：氯化 2,4-二甲基重氮苯

（2）偶氮化合物

这类化合物以—N＝N—两端连接芳基为特征，结构命名以"azo（偶氮）"为词头。构词采用如下方法：

azo-接芳烃全词 ＝ 偶氮芳烃

例如：

　　　　　英文名：*p*-aminoazobenzene
　　　　　中译名：对氨基偶氮苯

1.7.4　其他含氮有机物

① 结构为 R—NO_2 者，采用如下方法：

nitro-接母体全词 ＝ 硝基某烃

如 3-硝基甲苯的英文名为：3-nitrotoluene。

② 结构为 R—NO 者，采用如下方法：

nitroso-接母体全词 ＝ 亚硝基化合物

如对亚硝基-*N*,*N*-二甲基苯胺的英文名为：*p*-nitroso-*N*,*N*-dimethylaniline。

③ 结构为 R-ONO_2，采用如下方法：

R 基＋nitrate ＝ 硝酸某酯

如硝酸乙酯的英文名为：ethyl nitrate。

④ 结构 R-ONO 称为亚硝酸某酯，采用如下方法：

R 基＋nitrite ＝ 亚硝酸某酯

如亚硝酸甲酯的英文名为:methyl nitrite。
⑤ 氰化物的命名与无机无氧酸盐一样,采用如下方法:

$$\text{金属元素} + \text{cyanide} = \text{氰化物}$$

如氰化钾的英文名为:potassium cyanide。

碳链连接—CN 的有机化合物叫做"腈",已经在醛酮中讲过,即烃全词连-nitrile。如乙腈的英文名为:ethane nitrile。此外,还有其他一些命名。

基团结构	英文前缀	中译名
—COOCH$_3$	methoxycarbonyl-	甲氧碳酰基
—COOC$_6$H$_5$	phenoxycarbonyl-	苯氧碳酰基
—COONH$_2$	carbamoyl-	氨基甲酰基
—COCl	chloroformyl-	氯甲酰基
—COCH$_3$	acetyl-(aceto-)	乙酰基
—CN	cyano-	氰基

Lesson 8　Sulfur Compounds（含硫化合物）

1.8.1　含氧有机化合物相应的硫化物

该类化合物的中文命名,事实上只需在其含氧有机化合物的官能团名称前加一个"硫"字,就是对照的硫化物。其类别英汉对照见表 1-3。

落实到具体物质时,既然含氧有机化合物除 IUPAC 系统命名外尚有其他命名,那么对应的含硫化合物也应如此。只要了解构词的变化法则,照搬公式即可。

首先介绍"thio-"作为"硫代"的词头和词中,它在负二价硫化物的 IUPAC 系统命名中最为常见。要注意,这个词头一定要和"thia-（硫杂）"区别开。

表 1-3　含氧有机化合物相应的硫化物英汉对照

含氧有机化合物			相应的硫化物		
结构	英文名	中译名	结构	英文名	中译名
ROH	alcohol	醇	RSH	thio-alcohol	硫醇
—OH	hydroxy	羟基	—SH	hydrosulfury hydrosulfo- thiohydroxy (mercapto-)	巯基
ROR′	ether	醚	RSR′	thio-ether	硫醚
RCOR′	ketone	酮	RCSR′	thio-ketone	硫酮
RCHO	aldehyde	醛	RCHS	thio-aldehyde	硫醛

续表

含氧有机化合物			相应的硫化物		
结构	英文名	中译名	结构	英文名	中译名
\C=O	carboxyl	羰基	\C=S	thiocarbonyl / -thione	硫羰基 / 硫酮
ArOH	phenol	酚	ArSH	thiophenol	硫酚
RCOOH	carboxylic acid	羧酸	R-C(=S)-OH	-thionic acid	硫羰酸
			R-C(=O)-SH	-thiolic acid	硫羟酸
			R-C(=S)-SH	-dithioacid	二硫代羧酸
RCOOR′	ester	酯	R-C(=S)-OR′	-thionalt	硫羰酸酯
			R-C(=O)-SR′	-thiolate	硫羟酸酯
			R-C(=S)-SR′	dithio-ate	二硫代酸酯
ROOR′	peroxide	过氧化物	RSSR′	disulfide	二硫化物

—SH 化合物的 IUPAC 构词法为：

烃全词接-thiol ＝ 某硫醇

普通命名法为：

烃基＋mercaptan ＝ 某硫醇

如乙硫醇的 IUPAC 名称为：ethanethiol，普通命名：ethyl mercaptan。

同理，二硫醇、三硫醇应该用-dithiol、-trithiol 做词尾，构词方法同上。

"巯基"的 IUPAC 词头、词中名称为：thiohydroxy-，hydrosulfuryl-，hydrosulfo-，普通命名：mercapto-。

如 2,3-二巯基丙醇的 IUPAC 名称为：2,3-dihydrosulfopropanol。二甲硫醚的 IUPAC 名称为：methylthiomethane，普通命名：dimethyl thioether (dimethyl sulfide)。

显然，硫醚的构词与醚基本一致，IUPAC 系统命名方法是：

把 alkoxy-换成 alkylthio-，即烃氧基变成烃硫基，词尾不变。

普通命名公式为：

变 ether 为 sulfide 或 thioether

1.8.2 不含氧的含硫有机化合物

这类化合物中的硫往往是高价，特别是六价态和四价态，因此，常被当作硫酸、亚硫酸的衍生物。构词可先从对比说起。

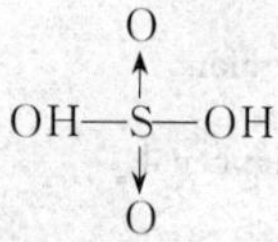

$$\text{OH}-\overset{\overset{O}{\uparrow}}{\underset{\underset{O}{\downarrow}}{S}}-\text{OH} \qquad \text{sulfuric acid 硫酸} \qquad \text{OH}-\overset{\overset{O}{\uparrow}}{S}-\text{OH} \qquad \text{sulfurous acid 亚硫酸}$$

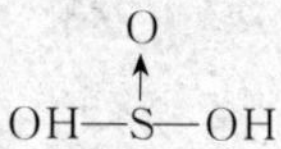

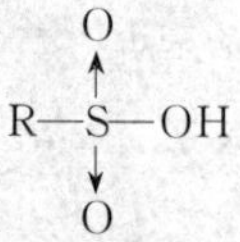

$$\text{R}-\overset{\overset{O}{\uparrow}}{\underset{\underset{O}{\downarrow}}{S}}-\text{OH} \qquad \text{sulfonic acid 磺酸} \qquad \text{R}-\overset{\overset{O}{\uparrow}}{S}-\text{OH} \qquad \text{sulfinic acid 亚磺酸}$$

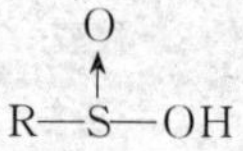

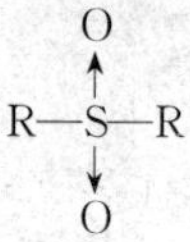

$$\text{R}-\overset{\overset{O}{\uparrow}}{\underset{\underset{O}{\downarrow}}{S}}-\text{R} \qquad \text{sulfone 砜} \qquad \text{R}-\overset{\overset{O}{\uparrow}}{S}-\text{R} \qquad \text{sulfoxide 亚砜}$$

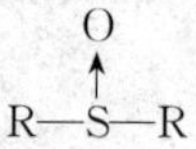

左右两边分别是两组六价、四价含硫有机化合物的基本词尾词汇。当用作词头基团时，—SO₃H 叫磺基（sulfo-）、—SO₂H 叫亚磺基（suifonyl-）。还有一个正二价态的硫—SOH，做主官能团叫次磺酸（-sulfenic acid），做词头叫次磺基（sulfeno-）。

上述硫在价态上不能由氧替换，故应专作记忆。但是其中，氧的位置处处可为硫替换。硫替换氧时为负二价，构词公式前面已讲述。

如硫代硫酸的英文名为：thiosulfuric acid。

从硫酸开始依次替换以便记忆掌握。由于它是二元酸，因此可以生成酸式盐与正盐。同理，与醇可生成一元酯与二元酯。含氧酸盐、酯的构词公式跟羧酸的构词法一致，词尾采用-ate。

例如：

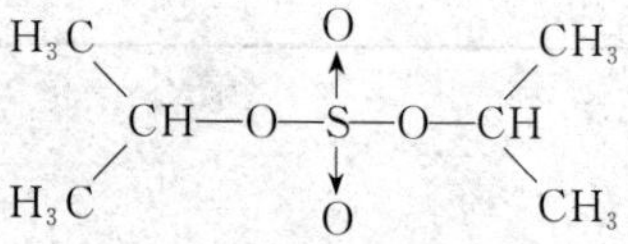

英文名：diisopropyl sulfate
中译名：硫酸二异丙酯

这里没有必要把异丙基写成 1-methylethyl。

若 R—取代了硫酸中的一个羟基，RSO₃H 结构中译名为烃磺酸。构词法为：

<u>烃全词接-sulfonic acid ＝ 某磺酸</u>

如：

英文名：2-propanesulfonic acid
中译名：2-丙烷基磺酸

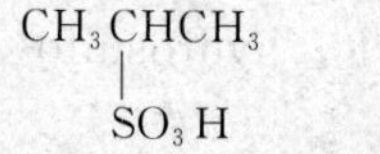

英文名：p-toluenesulfonic acid
中译名：对甲苯磺酸

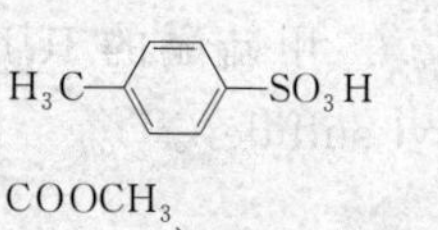

英文名：methyl 3-sulfobenzoate
中译名：3-磺基苯甲酸甲酯
（以—SO₃H 为前缀）

当然，磺酸也有它对应的盐和酯，其构词法为：

<u>金属全词（烃基）＋-sulfonate ＝ 磺酸盐（酯）</u>

如苯磺酸甲酯的英文名为：methyl benzenesulfonate。1-羟基乙磺酸钠的英文名为：sodium 1-hydroxyethanesulfonate。

同理，磺酸也有自己的酰基 R—SO$_2$—，英文名称为 sulfonyl。磺酸的磺酰基，常与 sulfuryl 硫酰基混用。如甲磺酰氯的英文名为：methanesulfonyl chloride。硫酰氯的英文名为：sulfury chloride。

结构 R—SO$_2$NH$_2$ 的中译名为磺酰胺，采用方法：

$$R \text{ 全词接-sulfonamide} = 某磺酰胺$$

结构 R—SO$_2$—R′ 的中译名为砜，采用方法：

$$烃基＋烃基＋sulfono = 某砜$$

如：

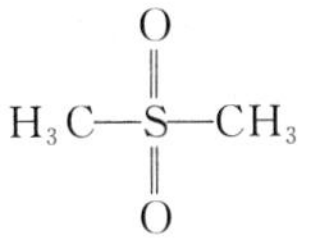

英文名：dimethyl sulfone
［(methylsulfonyl) methane］
中译名：二甲砜（甲磺酰甲烷）

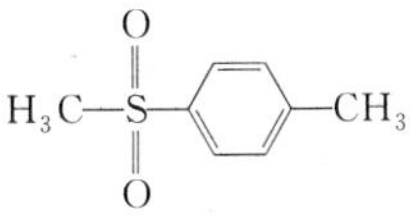

英文名：methyl *p*-tolyl sulfone
［4-methylsulfonyltoluene］
中译名：甲基对甲苯基砜（4-甲磺酰基甲苯）

Lesson 9   Phosphorus Compounds（含磷化合物）

和氧族比较，磷与氮同为第五主族，与硫同周期且相近，其价态变化也较多。

在有机化合物中，含磷的化合物有"磷、膦、鏻"三种译法，用汉语偏旁特征来表示磷元素的结构状态。其大致原则是：磷原子不直接与碳原子成键，用"磷"字；与碳原子直接成键用"膦"字；与碳原子共价直连且是配位正离子态者，用"鏻"字中译。其相应的酸、盐、酯等衍生物也都按规定选用中译的偏旁。所以，英文也就按各类变化分述如下。

1.9.1　PH$_3$ 与 NH$_3$ 比较

PH$_3$ 与 NH$_3$ 相似，有四种衍生物：伯膦（1°膦）RPH$_2$，英文名：primary phosphine；仲膦（2°膦）R$_2$PH，英文名：secondary phosphine；叔膦（3°膦）R$_3$P，英文名：tertiary phosphine；季鏻化合物［R$_4$P］$^+$X$^-$，英文名：quaternary phosphonfum compound。

伯、仲、叔膦的构词与胺类似，采用如下方法：

$$烃基依次接\text{-phosphine} = 某（基）膦$$

如三苯膦的英文名为：triphenylphosphine。

PH$_4^+$ 与 R$_4$P$^+$ 都用-phosphonium（鏻），构词方法：

$$烃基依次接\text{-phosphonium}＋阴离子 = 鏻化物$$

如氯化苯甲基三苯基磷的英文名为：benzyltriphenylphosphonium chloride。

1.9.2 五价磷酸及其衍生物

英文名：(ortho) phosphoric acid
中译名：（正）磷酸

这个酸的酯或盐按照类似硫酸的构词方法：

<u>醇烃基（金属）+ -phosphate ＝ 磷酸酯（盐）</u>

分情况命名实例如下：

英文名：isopropyl (dihydrogen) phosphate
中译名：磷酸（二氢）异丙酯

英文名：dimethyl (hydrogen) phosphate
中译名：磷酸（氢）二乙酯

英文名：trimethyl phosphate
中译名：磷酸三甲酯

当磷酸的羟基被烃基取代时，按磷碳相连原则，中译名为"膦"，保留酸字，构词方法为：

<u>烃基依次接-phosphonic acid ＝ 某膦酸</u>

例如：

英文名：1-propylphosphonic acid
中译名：1-丙膦酸

英文名：ethylmethylphosphonic acid
中译名：甲基乙基膦酸

其盐与酯只需将词尾-ic acid 变为-ate 即可，如：

英文名：diethyl methylphosphonate
中译名：甲膦酸二乙酯

1.9.3 三价磷酸及其衍生物

亚磷酸 H_3PO_3 的英文名为：phosphorous acid。正酸变亚酸是将词尾-ic acid 变成-ous acid。亚磷酸酯或盐是将词尾-ate 变为-ite。采用公式：

<u>醇烃基＋phosphite ＝ 亚磷酸酯（盐）</u>

如亚磷酸三乙酯 $(C_2H_5O)_3P$ 的英文名为：triethyl phosphite。

同理，$RP(OH)_2$ 及 R_2POH 结构都中译为亚磷酸，以-phosphonous acid 为词尾。

如乙亚磷酸 $C_2H_5P(OH)_2$ 的英文名为：ethylphosphonous acid。

也同理，将词尾-ous acid 变成-ite，即得到了对应酯。

如乙亚磷酸二甲酯 $C_2H_5P(OCH_3)_2$ 的英文名为：dimethyl ethylphosphonite。

Part 2
General Chemistry in Light Chemical Engineering

Lesson 1   Organic Chemistry

2.1.1   The Origins of Organic Chemistry

The term organic has acquired a variety of meanings during the 20th century, apart from its traditional usage to describe the branch of chemistry you are beginning to study here. For instance, some produce in the local store is described as organically grown, or we talk about organic foods. This label often means that a vegetable or fruit is grown under conditions where nutrients made by synthetic means are not added to the soil and that the nutrients come from compost instead. Another use of the word organic was popularized by Frank Lloyd Wright, who coined the term organic architecture to describe designs in which the exterior space of a building would blend with the surrounding environment of nature in a harmonious way. The word organic often attaches some notion of life or naturalness to an object.

In this way, the term organic chemistry was applied during the early 1800s to the study of substances related to living, or once-living, organisms. Inorganic was used to designate those substances such as rocks and minerals that have no life. But as time passed, differentiating between substances that came from plants and animals and those that originated from mineral sources became more difficult. Instead, materials began to be defined on the basis of their physical properties. Then, it became more important whether materials have high or low melting points, are soluble or insoluble in water, or are liquid, solid or gas.

Using these criteria, chemists soon realize that compounds containing carbon share many of the same properties. So, after isolating and characterizing literally millions of substances over the last two centuries, we have come to define organic chemistry as the chemistry of substances containing carbon. Almost all containing carbon falls into the classification of organic chemicals.

The field of organic chemistry is very broad, encompassing structures and reactions of several million molecules. It is an important area for study and research for many reasons, two of which will be mentioned here. First, the structures of organic molecules are now very well understood, and new types of molecules are readily prepared by rational approaches. This ability to make new substances with predictable structures and properties is one of the most important aspects that separate organic chemistry from many other branches of chemistry, physics and biology. The development of new drugs and materials for industrial and biomedical applications continues to require trained chemists who are interested in chemical synthesis. Second, the essential processes of biology are chemical reaction of organic substances. To discern how life processes occur, we must understand the fundamentals of organic chemistry.

2.1.2 The Subdisciplines of Organic Chemistry

In the study of organic chemistry, you will encounter several subdisciplines. Three areas that have been particularly important during development of the field and are still important today are physical organic chemistry, bioorganic chemistry and organometallic chemistry.

Physical organic chemistry encompasses studies of the pathways of chemical reactions of organic compounds and a systematic examination of all the variables involved in such processes. The structure and bonding of organic molecules as well as chemical reactions and their mechanisms (how reactions occur) fit into this broad subdiscipline.

Bioorganic chemistry concentrates on those compounds obtained from nature what we call natural products. Researches in this area focus on how such molecules are made and how they are involved in biochemical transformations. Bioorganic chemistry, in some respects, overlap with the study of molecular biology.

Organometallic chemistry is the study of compounds in which carbon is bonded to a metal atom or ion. Although many aspects of organometallic chemistry research focus on the properties of the metal ions themselves, and thus might be better classified as inorganic chemistry, the use of metal carbon bonded substances to prepare new materials constitutes an enormously large subdiscipline. Because many organometallic compounds are extremely reactive toward oxygen and water, the discovery of an organometallic species in nature, vitamin B_{12}, made us realize that compounds having a metal-carbon bond are sometimes quite stable and carry out chemical reactions that are otherwise difficult to do in a selective way.

Another way to classify subdisciplines of organic chemistry is to look at what organic chemists "do". For example, those interested in preparing new substances are said to be involved in synthetic organic chemistry. Those whose attention is focused on characterizing molecular structures study analytical organic chemistry or organic spectroscopy. Chemists specializing in molecules with very high, molecular mass, especially those substances that have regularly repeating structural motifs, are organic polymer chemists. The use of light to carry out chemical reactions of organic compounds has led to the growth of the field of organic photochemistry.

2.1.3　The Applications of Computer in Organic Chemistry

A familiar arrangement of the sciences places chemistry between physics which is highly mathematical, and biology which is highly descriptive. Among chemistry's subdisciplines, organic chemistry is less mathematical than descriptive in that it emphasizes the qualitative aspects of molecular structure, reaction and synthesis. The earliest applications of computer to chemistry took advantage of the "number crunching" power of mainframes to analyze data and to perform calculations concerned with the more quantitative aspects of bonding theory. More recently, organic chemists have found the graphics capabilities of minicomputer, workstations and personal computers to be well-suited to visualizing a molecule as a three-dimensional object and assessing its ability to interact with another molecule. Given a biomolecule of known structure, a protein, for example, and a drug that acts on it, molecular-modeling software can evaluate the various ways in which the two may fit together.

2.1.4　Challenges and Opportunities

After World War II, the growth of organic chemistry occurred rapidly, as the field assimilated new results from many other disciplines. The early 1950s, in particular, saw the birth of many subfields of chemistry, including modern organometallic chemistry, beginning with the synthesis and characterization of ferrocene; molecular biology, with the elucidation of the structure of DNA; structural biochemistry, with the description of bonding in proteins and the first crystal structure of a protein; and modern instrumental analysis of organic structures, especially the development of nuclear magnetic resonance (NMR) spectroscopy. All of these propitious discoveries opened many new avenues of chemical research that were incorporated into the growing field of organic chemistry.

A major contribution to the growth of organic chemistry during this century has been the accessibility of cheap starting materials. Petroleum and natural gas provide the building blocks for the construction of larger molecules. Many drugs, plastics, synthetic fibers, films, and elastomers are made from the organic chemicals obtained from petroleum. As we enter an age of inadequate and shrinking supplies, the use to which we put petroleum looms large in determining the kind of society we will have. Alternative sources of energy, especially for transportation, will allow a greater fraction of the limited petroleum available to be converted to petrochemicals instead of being burned in automobile engines. At a more fundamental level, scientists in the chemical industry are trying to devise ways to use carbon dioxide as a carbon source in the building block molecules.

New Words and Expressions

bioorganic ['baɪəʊɔːˈgænɪk] adj. 生物有机的	plastic ['plæstɪk] n. 塑料制品; adj. 塑料的
bonding ['bɒndɪŋ] n. 键合,偶合,粘接,焊接	protein ['prəʊtiːn] n. 蛋白质; adj. 蛋白质的
compost ['kɒmpɒst] n. 混合肥料,堆肥	quantitative ['kwɒntətetɪv] adj. 定量的,数量的
elastomer [ɪ'læstəmə] n. 人造橡胶,弹性体	shrink [ʃrɪŋk] v. 收缩,使收缩,缩短
elucidation [ɪˌluːsɪ'deɪʃn] n. 说明,阐明	spectroscopy [spek'trɒskəpɪ] n. 光谱学,波谱学
film [fɪlm] n. 薄膜; v. 在……上覆以薄膜	subdiscipline [sʌb'dɪsɪplɪn] n. 学科分支
nutrient ['njuːtrɪənt] n. 营养物,营养剂	synthetic [sɪn'θetɪk] adj. 合成的,人造的
organic [ɔː'gænɪk] n. 有机的,器官的	fiber ['faɪbə] n. 纤维
organism ['ɔːgənɪzəm] n. 生物体,有机体	nuclear magnetic resonance (NMR) 核磁共振
organometallic [ɔːˌgænəʊmɪ'tælɪk] adj. 有机金属的	synthetic fiber 合成纤维
overlap [ˌəʊvə'læp] v. 与……交叠	motif [məʊ'tiːf] n. 结构基元,结构域
propitious [prə'pɪʃəs] adj. 有利的	loom [luːm] v. (危险) 阴森森地逼近

Lesson 2 Polymer Chemistry

Many of the terms and definitions used in polymer chemistry are not encountered in conventional chemical textbooks, and for this reason the following summary of terminology is given. Some of these definitions will seem fairly obvious, but others will need explanation.

2.2.1 Monomers

A monomer is any substance that can be converted into a polymer. For example, ethylene is a monomer that can be polymerized to polyethylene.

$$n\text{CH}_2=\text{CH}_2 \longrightarrow \pm\text{CH}_2-\text{CH}_2\pm_n \tag{2-1}$$

An amino acid is a monomer which, by loss of water, can polymerize to give polypeptides. The term monomer is used very loosely, sometimes it applies to dimers or trimers if they, themselves, can undergo further polymerization.

$$n\,\text{H}_2\text{N}-\underset{\underset{\text{H}}{|}}{\overset{\overset{\text{R}}{|}}{\text{C}}}-\overset{\overset{\text{O}}{\|}}{\text{C}}-\text{OH} \xrightarrow{-\text{H}_2\text{O}} \left[-\text{N}-\underset{\underset{\text{H}}{|}}{\overset{\overset{\text{R}}{|}}{\text{C}}}-\overset{\overset{\text{O}}{\|}}{\text{C}}-\right]_n \tag{2-2}$$

2.2.2 Polymers

The term polymer is used to describe high-molecular-mass substances. However, this is a very broad definition, and in practice it is convenient to divide polymers into subcategories according to their molecular mass and structure. Although there is no general agreement on this point, in this book we will consider low polymers to have molecular mass below about 10,000 to 20,000 and high polymers to have molecular mass between 20,000 and several million. Obviously, this is a rather arbitrary dividing line, and a better definition might be based on the number of repeating units in the structure. For example, since polymer properties become almost independent of molecular mass when more than 1,000 to 2,000 repeating units are present, this point could also constitute a satisfactory dividing line between low and high polymers.

2.2.3 Linear Polymers

A linear polymer consists of along chain of skeletal atoms to which are attached the substituent groups. Polyethylene (**1**) is one of the simplest examples. Linear polymers are usually soluble in some solvent, and in the solid state at normal temperatures they exist as elastomers, flexible materials, or glasslike thermoplastics. In addition to polyethylene, typical liner-type polymers include poly (vinyl chloride) or PVC (**2**), poly (methyl methacrylate) (also known as PMMA, Lucite, Plexiglass or Perspex) (**3**), polyacrylonitrile (Orlon or Creslan) (**4**), and nylon 66 (**5**).

$$\left[\begin{array}{cc} H & H \\ | & | \\ C-C \\ | & | \\ H & H \end{array}\right]_n \quad \left[\begin{array}{cc} H & Cl \\ | & | \\ C-C \\ | & | \\ H & H \end{array}\right]_n \quad \left[\begin{array}{cc} & O-CH_3 \\ & | \\ H & C=O \\ | & | \\ C-C \\ | & | \\ H & CH_3 \end{array}\right]_n \quad \left[\begin{array}{cc} H & C\equiv N \\ | & | \\ C-C \\ | & | \\ H & H \end{array}\right]_n$$

(1)   (2)   (3)   (4)

$$\left[\begin{array}{c} H \\ | \\ N-(CH_2)_6-N-C-(CH_2)_4-C \\ \quad\quad\quad\quad\quad | \quad \| \quad\quad\quad\quad \| \\ \quad\quad\quad\quad\quad H \quad O \quad\quad\quad\quad\quad O \end{array}\right]_n$$

(5)

2.2.4 Branched Polymers

A branched polymer can be visualized as a linear polymer with branches of the same basic structure as the main chain. A branched polymer structure is illustrated in Figure 2-1. Branched polymers are often soluble in the same solvents as the corresponding linear polymer. In fact, they resemble linear polymer in many of their properties. However, they can sometimes be distinguished from linear polymers by their lower tendency to crystallize or by their different solution viscosity or light-scattering behavior. Heavily branched polymers may swell in certain liquids without dissolving completely.

Figure 2-1 Branched polymer

2.2.5 Crosslinked Polymers

A crosslinked or network polymer is one in which chemical linkages exist between the chains, as illustrated in Figure 2-2. Such materials are usually swelled by "solvents", but they do not dissolve. In fact, this insolubility can be used as a cautious criterion of a crosslinked structure. Actually, the amount by which the polymer is swelled by a liquid depends on the density of crosslinking, the more crosslinks present, the smaller is the amount of swelling. If the degree of crosslinking is high enough, the material may be a rigid, high-melting, unswellable solid, such as diamond. Light crosslinking of chains favors the formation of rubbery elastomeric properties.

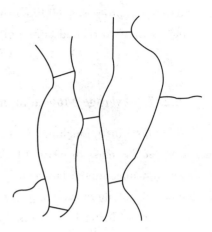

Figure 2-2 Crosslinked polymer

2.2.6 Copolymers

A copolymer is a polymer made from two or more different monomers. For example, if styrene and acrylonitrile are allowed to polymerize in the same reaction vessel, a copolymer will be formed which contains both styrene and acrylonitrile residues. Many commercial synthetic polymers are copolymers. It should be noted that the sequence of monomer units along a copolymer chain can vary according to the method and mechanism of synthesis. Three different types of sequencing arrangements are commonly found.

$$n\,\underset{CH=CH_2}{\bigcirc} + n\,\underset{CH=CH_2}{C\equiv N} \longrightarrow \left[\underset{H}{\underset{|}{C}}-CH_2-\underset{|}{\underset{C\equiv N}{CH}}-CH_2\right]_n$$

copolymer   (2-3)

(1) Random copolymers. In random copolymers, no definite sequence of monomer units exists. A copolymer of monomers A and B might be depicted by the arrangement shown in (**6**). Random copolymers are often formed when olefin-type monomers copolymerize by free-radical-type processes. The properties of random copolymers are usually quite different from those of the related homopolymers.

$$-A-B-B-B-A-A-B-A-A-A-A-B-A-B-B-B-$$

(6)

(2) Alternating copolymers. As the name implies, alternating copolymers contain a regular alternating sequence of two monomer units (**7**). Olefin polymerizations that take place through ionic-type mechanisms can yield copolymers of this type. Again, the properties of the copolymer usually differ markedly from those of the two related homopolymers.

$$-A-B-A-B-A-B-A-B-$$

(7)

(3) Block copolymers. Block copolymers contain a block of one monomer connected to a block of another, as illustrated in sequence (8). Block copolymers are often formed by ionic polymerization processes. Unlike other copolymers, they retain many of the physical characteristics of the two related homopolymers.

$$-A-A-A-A-A-A-A-A-B-B-B-B-B-B-B$$
(8)

2.2.7 Average Molecular mass and Distributions

A major distinguishing feature of high polymers is their enormous molecular mass. Molecular mass of 20,000 Daltons are routine, and values as high as 2,000,000 Daltons are not uncommon. However, unlike small molecules such as benzene or chloroform, or biological polymers like enzymes, a sample of a synthetic polymer has no single, fixed molecular mass. Instead, there is a distribution of different molecular mass in the same sample of material (Figure 2-3). For this reason, it is necessary to speak of average molecular mass rather than a single defining value.

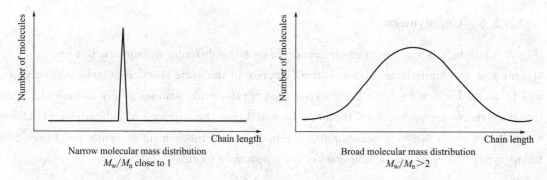

Figure 2-3 Molecular mass distributions

Several different types of average molecular mass are used in polymer chemistry, the most important of which are known as number average, $\overline{M}_n$, and mass average, $\overline{M}_w$, values. They are defined as shown in equations (2-4) and (2-5):

$$\overline{M}_n = \sum_i X_i M_i = \frac{\sum_i N_i M_i}{\sum_i N_i} \tag{2-4}$$

$$\overline{M}_w = \sum_i W_i M_i = \frac{\sum_i w_i M_i}{\sum_i w_i} = \frac{\sum_i N_i M_i^2}{\sum_i N_i M_i} \tag{2-5}$$

where N_i is the number of molecules of molecular mass M_i, and X_i is the number fraction or mole fraction of molecules having molecular mass M_i, and where w_i is the mass of molecules of molecular mass M_i, and W_i is the mass fraction of molecules with molecular mass M_i.

In general, $\overline{M}_w$ values are higher than $\overline{M}_n$ because the calculations for $\overline{M}_w$ give more emphasis to the larger molecules, while $\overline{M}_n$ calculations give equal emphasis to all molecules.

The fraction $\overline{M}_w/\overline{M}_n$ (called the polydispersity) is a measure of the molecular mass distribution. If the value of $\overline{M}_w/\overline{M}_n$ is close to 1 (1.01 or 1.02 for instance), the distribution is very narrow. If it is, say, 2 or higher, the distribution is considered to be very broad. The molecular-mass distribution affects several important polymer properties. For example, polymers with very broad distributions are less prone to crystallize than their narrow distribution counterparts, and they often have lower solidification temperatures. The shorter chains plasticize the bulk material and make it softer. Thus, together with the glass transition temperature, T_g, and the crystalline melting temperature, T_m (see below), $\overline{M}_w/\overline{M}_n$ is a crucial characteristic of any synthetic polymer.

2.2.8 Themoplastics

Basically, a thermoplastic is any materiel that softens when it is heated. However, the term is commonly used to describe a substance that passes through a definite sequence of property changes as its temperature is raised. In Figure 2-4 the thermoplastic characteristics of amorphous, crystalline and liquid crystalline polymers are compered. An amorphous polymeric material contains randomly entangled chains. A microcrystalline (usually abbreviated to "crystalline") material contains domains in which the polymer chains are packed in an ordered array. These "crystalline" domains are embedded in an amorphous polymer matrix.

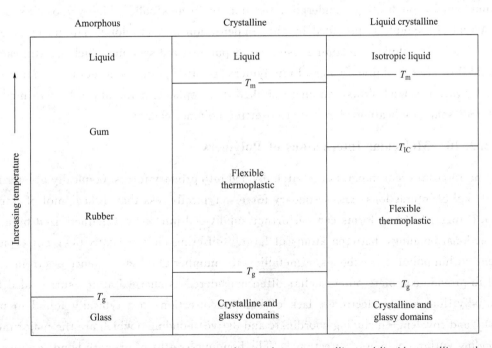

Figure 2-4 Comparison of the thermal behavior of amorphous, crystalline and liquid crystalline polymers

Both amorphous and crystalline thermoplastics are glasses at low temperatures, and both change from a glass to a rubbery elastomer or flexible plastic as the temperature is raised. This change from glass to elastomer usually takes place over a fairly narrow temperature range (2-5°C), and this transition point is known as the glass transition temperature

(T_g). For many polymers, the glass transition temperature is the most important characterization feature. It can be compared to the characteristic melting point of a low-molecular-mass compound, although care should be taken to remember that T_g is definitely not a melting temperature in the accepted sense of the word. It is more, a measure of the ease of torsion of the backbone bonds rather than of the ease of separation of the molecules.

At temperatures above T_g, amorphous polymers behave in a different manner from crystalline polymers. As the temperature of an amorphous polymer is raised, the hard rubbery phase gradually gives way to a soft, extensible elastomeric phase, then to a gum, and finally to a liquid. No sharp transition occurs from one phase to the other, and only a gradual change in properties is perceptible.

Crystalline polymers, on the other hand, retain their rubbery elastomeric or flexible properties above the glass transition, until the temperature reaches the melting temperature (T_m). At this point, the material liquefies. At the same time, melting is accompanied by a loss of the optical birefringence and crystalline X-ray diffraction effects that are characteristic of the crystalline state.

2.2.9 Thermosetting Resin

The term thermosetting polymer refers to a range of systems which exist initially as liquids but which, on beating, undergo a reaction to form a solid, highly crosslinked matrix. A typical example is provided by the condensation of methylol melamine to give the hard, tough, crosslinked melamine resin. Partly polymerized systems which are still capable of liquid flow are called prepolymers. Prepolymers are often preferred as emitting materials in technology. In practical terms, an uncrosslinked thermoplastic material can be reformed into a different shape by heating, but a thermosetting polymer cannot.

2.2.10 Molecular Interactions of Polymers

The forces present in nature are often divided into primary forces (typically greeter than 50kcal/mol of interactions) and secondary forces (typically less than 10kcal/mol of interactions). Primary bonding forms can be further subdivided into ionic (characterized by a lack of directional bonding, between atoms of largely differing electronegativities; not typically present within polymer backbones), metallic (the number of outer, valence electrons is too small to provide complete outer shells; often considered as charged atoms surrounded by a potentially fluid sea of electrons; lack of bonding direction; not typically found in polymers), and covalent (including coordinate and dative) bonding (which are the major means of bonding within polymers, directional). The bonding lengths of primary bonds are usually about 0.09-0.2nm with the carbon-carbon bond length being about 0.15-0.16nm.

Secondary forces, frequently called van der Waals forces, since they are the forces responsible for the van der Waals correction to the ideal gas relationships, are of longer distance in interaction generally having significant interaction between 0.25nm and 0.5nm. The force of these interactions is inversely proportional to some power of r, generally 2 or greater [force ∝ 1/ (distance)^r] and thus is quite dependent on the distance between the interacting molecules. Thus, many physical properties of polymers are indeed quite dependent on both the conformation (arrangements related to rotation about single bonds) and configuration (arrangements related to the actual chemical bonding about a given atom) since both affect the proximity one chain can have relative to another. Thus, amorphous polypropylene is more flexible than crystalline polypropylene [compare liner polymers (a) and (b) in Figure 2-5].

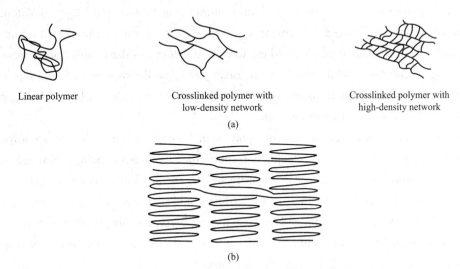

Figure 2-5 Representation of an amorphous polymer and representation of folded polymer chains in polymer crystals

Atoms in individual polymer molecules are thus joined to each other by relatively strong covalent bonds. The bond energies of the carbon-carbon bonds are on the order of 80 to 90kcal/mol. Further polymer molecules, like all other molecules, are attracted to each other (and for long-chain polymer chains even between segments of the same chain) by intermolecular, secondary forces.

These intermolecular forces are also responsible for the increase in boiling points within a homologous series such as the alkanes, for the higher than expected boiling points of polar organic molecules such as alkyl chlorides, and for the abnormally high boiling points of alcohols, amines and, amides. While the forces responsible for these increases in boiling points are all called van der Waals forces, these forces are subclassified in accordance with their sources and intensities. Secondary, intermolecular forces include London dispersion forces, induced permanent forces, and dipolar forces, including hydrogen bonding.

Nonpolar molecules such as ethane [$H(CH_2)_2H$] and polyethylene are attracted to each other by weak London or dispersion forces resulting from induced dipole-dipole interaction. The temporary or transient dipoles in ethane or along the polyethylene chain are due to instantaneous fluctuations in the density of the electron clouds. The energy range of these forces is about 2kcal per unit in the nonpolar and polar polymers alike, and this force is independent of temperature. These London forces are typically the major forces present between chains in largely nonpolar polymers present in elastomers and soft plastics.

It is of interest to note that methane, ethane, and ethylene are all gases; hexane, octane, and nonane are all liquids (at room conditions); while polyethylene is a waxy solid. This trend is primarily due to both an increase in mass per molecule and to an increase in the London forces per molecule as the chain length increases.

Polar molecules such as ethyl chloride ($H_3C—CH_2Cl$) and poly (vinyl chloride), $[\!-\!CH_2—CHCl\!-\!]_n$, PVC, PAN see Figure 2-6] are attracted to each other by dipole-dipole interactions resulting from the electrostatic attraction of a chlorine atom in one molecule to a hydrogen atom in another molecule. Since this dipole-dipole interaction, which ranges from 2 to 6kcal/mol repeat unit in the molecule, is temperature dependent, these forces are reduced as the temperature is increased in the processing of polymers. While the London forces are typically weaker than the dipole-dipole forces, they are also present in polar compounds, such as ethyl chloride and PVC. These dipole-dipole forces are characteristic of many plastics.

Strong polar molecules such as ethanol, poly (vinyl alcohol), and cellulose are attracted to each other by a special type of dipole-dipole interaction called hydrogen bonding, in which the oxygen or nitrogen atoms in one molecule are attracted to the hydrogen atoms in another molecule. These are the strongest of the intermolecular forces and may have energies as high as 10kcal/mol repeat unit. Intermolecular hydrogen bonds are usually present in fibers, such as cotton, wool, silk, nylon, acrylon, polyester, and polyurethane. Intramolecular hydrogen bonds are responsible for the helices observed in starch and globular proteins.

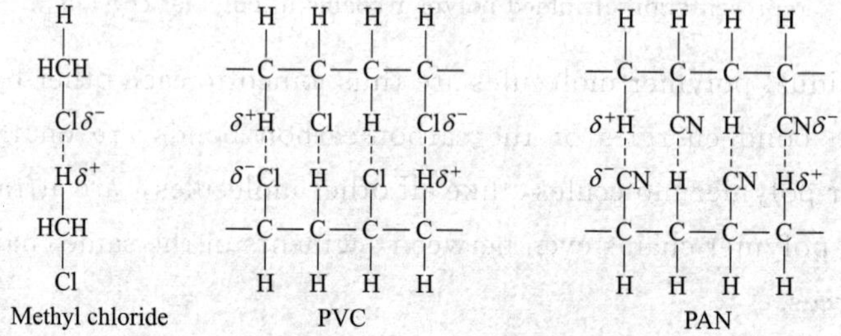

Figure 2-6 Typical dipole-dipole interaction between molecules of methyl chloride end segments of chains of poly (vinyl chloride) and polyacrylonitrile

It is important to note that the high melting point of nylon 66 (265℃) is the result of a combination of London, dipole-dipole, and hydrogen bonding forces between the polyamide chains (see Figure 2-7). The hydrogen bonds are decreased when the hydrogen atoms in the amide groups in nylon are replaced by methyl groups and when the hydroxyl groups in cellulose are esterified.

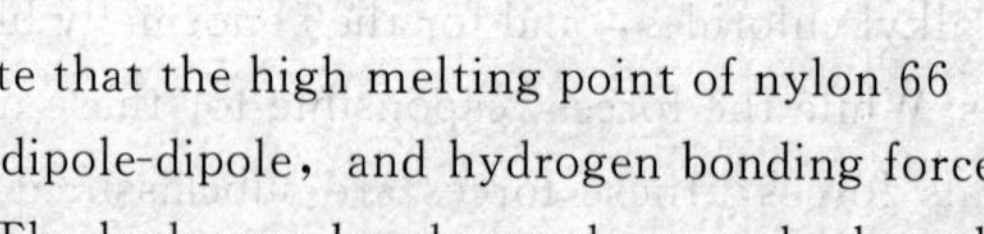

Figure 2-7 Typical hydrogen bonding between hydrogen and oxygen or nitrogen atoms in nylon 66

The flexibility of amorphous polymers is reduced drastically when they are cooled below a characteristic transition temperature called the glass transition temperature (T_g). At temperature below T_g, there is no segmental motion and any dimensional changes in the polymer chain are the result of temporary distortions of the primary valence bonds. Amorphous plastics perform best below T_g, but elastomers must be used above the brittle point, or T_g.

The flexibility of amorphous polymers above the glassy state is dependent on a wriggling type of segment motion in the polymer chains. This flexibility is increased when many methylene groups (CH_2) or oxygen atoms are present between stiffening groups is the chain. Thus, the flexibility of aliphatic polyesters usually increases as m is increased.

$$\{(CH_2)_m-O-\underset{\underset{O}{\|}}{C}-(CH_2)_m-\underset{\underset{O}{\|}}{C}-O\}_n$$

Aliphatic Polyester

In contrast, the flexibility of amorphous polymers above the glassy state is decreased when stiffening groups such as,

p-Phenylene   Amide   Sulfone   Carbonyl

are present in the polymer backbone. Thus, poly (ethylene terephthalate) is stiffer and higher melting point than poly (ethylene adipate) and the former is stiffer than poly (butylene terephthalate) because of the presence of fewer methylene groups between the stiffening groups.

Poly(ethylene adipate)   Poly(ethylene terephthalate)

The melting point (T_m) is called the first-order transition temperature, and T_g is sometimes called the second-order transition temperature. The values for T_m are usually 33% to 100% greater than T_g, and symmetrical polymers like HDPE exhibit the greatest difference between T_m and T_g.

As shown in Figure 2-8, values for both T_g and T_m are observed as endothermic transitions in calorimetric measurements, such as differential thermal analysis (DTA) or differential scanning calorimetry (DSC). It is important to note that since the values observed for T_g are dependent on the test method and on time, the values obtained by different techniques may vary by a few degrees.

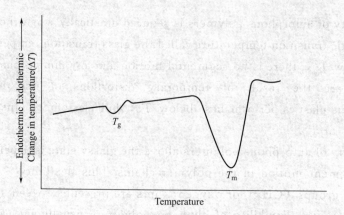

Figure 2-8　A Typical DTA thermogram of a polymer

New words and expression

monomer ['mɒnəmə] n. 单体	chloroform ['klɒrəfɔːm] n. 氯仿,三氯甲烷
polyethylene [,pɒlɪ'eθəliːn] n. 聚乙烯	amorphous [ə'mɔːfəs] adj. 无定形的
polypeptide [,pɒlɪ'peptaɪd] n. 多肽	birefringence [,baɪrɪ'frɪndʒəns] n. 双折射
dimer ['daɪmə] n. 二聚体,二聚物	thermosetting ['θɜːməʊ,setɪŋ] adj. 热固性的
trimer ['traɪmə] n. 三聚体,三聚物	alternating copolymer 交替共聚物
thermoplastics [θɜːmə'plæstɪks] n. 热塑性塑料	block copolymer 嵌段共聚物
polyacrylonitrile [,pɒlɪ'ækrələʊ'naɪtrɪl] n. 聚丙烯腈	number fraction 质量分数
copolymer [kəʊ'pɒlɪmə] n. 共聚物	mole fraction 摩尔分数
matrix ['meɪtrɪks] n. 基体,基质	synthetic polymer 合成聚合物
melamine ['meləmiːn] n. 三聚氰胺,蜜胺	random copolymer 无规共聚物
branched polymer 支化聚合物,支链聚合物	olefin-type monomer 烯烃类的单体
crosslinked polymer 交联聚合物	free-radical-type process 自由基(游离基)型反应
styrene ['staɪriːn] adj. 苯乙烯	polydispersity 多分散性
acrylonitrile [ækrələʊ'naɪtrɪl] n. 丙烯腈	methylol melamine 羟甲基三聚氰胺
homopolymer [həʊmə'pɒlɪmə] n. 均聚物	prepolymer 预聚物,预聚体
Dalton ['dɔːltən] 道尔顿(质量单位)	intramolecular [,ɪntrəmə'lekjʊlə] adj. 分子内的
conformation [,kɒnfɔː'meɪʃn] n. 构象	polyurethane [,pɒlɪ'jʊərəθeɪn] n. 聚氨酯,聚氨基甲酸酯
homologous series 同系列	melt viscosity 熔体黏度
induced permanent force 诱导力	poly(ethylene terephthalate)　聚对苯二甲酸乙二酯
dipoler force 偶极力	poly(ethylene adipate)　聚己二酸乙二酯
London force 伦敦力,色散力	poly(butylene terephthalate)　聚对苯二甲酸丁二酯
HDPE　(high density polyethylene)　高密度聚乙烯	

Lesson 3　Surface and Interface Chemistry

2.3.1　Surface Tension

It is well known that short-range van der Waals forces of attraction exist between molecules and are responsible for the existence of the liquid state. The phenomena of surface and interfacial tension are readily explained in terms of these forces. The molecules which are located within the bulk of a liquid are, on the average, subjected to equal forces of attraction in all directions, whereas those located, for example, at a liquid-air interface experience unbalanced attractive forces resulting in a net inward pull. As many molecules as possible will leave the liquid surface for the interior of the liquid; therefore, the surface will tend to contract spontaneously. For this reason droplets of liquid and bubbles of gas tend to attain a spherical shape.

The molecules at a surface have higher potential energies than those in the interior. This is because they interact more strongly with the molecules in the interior of the substance than they do with the widely spaced gas molecules above it. Work is therefore required to bring molecule from the interior to the surface.

Surface tension and the more fundamental quantity, surface free energy, fulfill an outstanding role in the physical chemistry of surfaces. The surface tension of a liquid is often defined as the force acting at right-angles to any line of unit length on the liquid surface. However, this definition (although appropriate in the case of liquid films, such as in foams) is somewhat misleading, since there is no elastic skin or tangential force as such at the surface of a pure liquid. It is more satisfactory to define surface tension and surface free energy as the work required to increase the area of a surface isothermally and reversibly by unit amount. Attractive force between molecules at surface and in the interior of a liquid is illustrated in Figure 2-9.

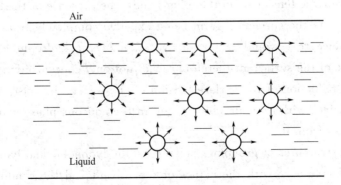

Figure 2-9　Attractive forces between molecules at surface and in the interior of a liquid

The same considerations apply to the interface between two immiscible liquids. Again there is an imbalance of intermolecular forces but of a lesser magnitude. Interfacial tensions usually lie between the individual surface tensions of the two liquids in question.

The above picture implies a static state of affairs. However, it must be appreciated that an apparently quiescent liquid surface is actually in a state of great turbulence on the molecular scale as a result of two-way traffic between the bulk of the liquid and the surface, and between the surface and vapour phase. The average lifetime of a molecule at the surface of a liquid is ca. 10^{-6} seconds.

2.3.2 Surfactant

Surface-active agents or surfactants are substances that, when present at low concentration in a system, have the property of adsorbing onto the surfaces or interfaces of the system and of altering to a marked degree the surface or interfacial free energies of those surfaces (or interfaces). They have a characteristic molecular structure consisting of a structural group that has very little attraction for the solvent, called the lyophobic group, together with a group that has strong attraction for the solvent called the lyophilic group. This is known as an amphipathic structure.

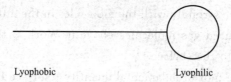

Lyophobic　　　　　Lyophilic

When a surface-active agent is dissolved in a solvent, the presence of the lyophobic group in the interior of the solvent may cause distortion of the solvent liquid structure, increasing the free energy of the system. In an aqueous solution of a surfactant this distortion of the water by lyophobic (hydrophobic) group of the surfactant increase the free energy of the system when it is dissolved, means that less work is needed to bring a surfactant molecule than a water molecule to the surface. The surfactant therefore concentrates at the surface. Since less work is now needed to bring molecules to the surface, the presence of the surfactant decreases the work needed to create unit area of surface (the surface free energy per unit area, or surface tension). On the other hand, the presence of the lyophilic (hydrophilic) group prevents the surfactant from being expelled completely from the solvent as a separate phase, since that would require dehydration of the hydrophilic group. The amphipathic structure of the surfactant therefore causes not only concentration of the surfactant at the surface and reduction of the surface tension of the water, but also orientation of the molecule at the surface with its hydrophilic group in the aqueous phase and its hydrophobic group oriented away from it.

The chemical structures of groupings suitable as the lyophobic and lyophilic portions of the surfactant molecule vary with the nature of the solvent and the conditions of use. In a highly polar solvent such as water, the lyophobic group may be a hydrocarbon or fluorocarbon or siloxane chain of proper length, whereas in a less polar solvent only some of these may be suitable (e. g. , fluorocarbon or siloxane chains in polypropylene glycol). In a polar solvent such as water, ionic or highly polar groups may act as lyophilic groups, whereas in a nonpolar solvent such as heptane they may act as lyophobic groups. As the temperature and use conditions (e. g. , presence of electrolyte or organic additives) vary, modifications in the structure of the lyophobic and lyophilic groups may become necessary to maintain surface activity at a suitable level. Thus, for surface activity in a particular system the surfactant molecule must have a chemical structure that is amphipathic in that solvent under the conditions of use.

The hydrophobic group is usually a long-chain hydrocarbon residue, and less often a halogenated or oxygenated hydrocarbon or siloxane chain; the hydrophilic group is an ionic or highly polar group depending on the nature of the hydrophilic group. Surfactants are classified as:

(1) Anionic. The surface-active portion of the molecule bears a negative charge, for example, $RCOO^-Na^+$ (soap), $RC_6H_4SO_3^-Na^+$ (alkylbenzene sulfonate).

(2) Cationic. The surface-active portion bears a positive charge, for example, $RNH_3^+Cl^-$ (salt of a long-chain amine), $RN(CH_3)_3^+Cl^-$ (quaternary ammonium chloride).

(3) Zwitterionic or amphoteric. Both positive and negative charges may be present in the surface-active portion, for example, $RN^+H_2CH_2COO^-$ (long-chain amino acid), $RN^+(CH_3)_2CH_2CH_2SO_3^-$ (sulfobetaine).

(4) Nonionic. The surface-active portion bears no apparent ionic charge, for e sample, $RCOOCH_2CHOHCH_2OH$ (monoglyceride of long-chain fatty acid), $RC_6H_4(OC_2H_4)_xOH$ (polyoxyeehylene alkylphenol).

Most natural surfaces are negatively charged. Therefore, if the surface is to be madehydrophobic (water-repellent) by use of a surfactant, then the best type of surfactant to use is a cationic. This type of surfactant will adsorb onto the surface with its positively charged hydrophilic head group oriented toward the negatively charged surface (because of electrostatic attraction) and its hydrophobic group oriented away from the surface, making the surface water-repellent. On the other hand, if the surface is to be made hydrophilic (water-wettable), then cationic surfactant should be avoided. If the surface should happen to be positively charged, however, then anionics will make it hydrophobic and should be avoided if the surface is to be made hydrophilic.

Nonionics adsorb onto surfaces with either the hydrophilic or the hydrophobic group oriented toward the surface, depending upon the nature of the surface. If polar groups capable of hydrogen bonding with the hydrophilic group of the surfactant are present on the surface, then the surfactant will probably be adsorbed with its hydrophilic group oriented toward the surface, making the surface more hydrophobic; if such groups are absent from the surface, then the surfactant will probably be oriented with its hydrophobic group toward the surface, making it more hydrophilic.

Zwittterionics, since they carry both positive and negative charges, can adsorb onto both negatively charged and positively charged surfaces without changing the charge of the surface significantly. On the other hand, the adsorption of a cationic onto a negatively charged surface reduces the charge on the surface and may even reverse it to a positive charge (if sufficient cationic is adsorbed). In similar fashion, the adsorption of an anionic surfactant onto a positively charged surface reduces its charge and may reverse it to a negative charge. The adsorption of a nonionic onto a surface generally does not affect its charge significantly, although the effective charge density may be reduced if the adsorbed layer is thick.

Differences in the nature of the hydrophobic groups are usually less pronounced than in the nature of the hydrophilic groups. Generally, they are long-chain hydrocarbon residues. However, they include such different structures as:

(1) straight-chain, long alkyl groups ($C_8 \sim C_{20}$);

(2) branched-chain, long alkyl groups ($C_8 \sim C_{20}$);

(3) long-chain ($C_8 \sim C_{15}$) alkylbenzene residues;

(4) alkylnaphthalene residues (C_3 and greater-length alkyl groups);

(5) high-molecular-mass propylene oxide polymers (polyoxypropylene glycol derivatives);

(6) long-chain perfluoroalkyl groups;

(7) polysiloxane groups;

(8) lignin derivatives.

2.3.3 Emulsion

An emulsion is a mixture of two immiscible (unblendable) substances. One substance (the dispersed phase) is dispersed in the other (the continuous phase). Emulsification is the process by which emulsions are prepared. It is probably the most versatile of surface active agents for practical applications and, as a result, has been extensively studied.

Two immiscible, pure liquids cannot form an emulsion. For a suspension of one liquid in another to be stable enough to be classified as an emulsion, a third component must be present to stabilize the system. The third component is called the emulsifying agent which is usually a surfactant.

Macroemulsions are of two types, based on the nature of the dispersed phase, oil-in-water (O/W) and water-in-oil (W/O). The oil-in-water type is a dispersion of a water-immiscible liquid or solution, always called the oil (O), regardless of its nature, in an aqueous phase (W). The oil is, in this case, the discontinuous (inner) phase; the aqueous phase is the continuous (outer) phase. The water-in-oil type is a dispersion of water or an aqueous solution (W) in a water-immiscible liquid (O). The type of emulsion formed by the water and the oil depends primarily on the nature of the emulsifying agent, to a minor extent, on the process used in preparing the emulsion and the relative proportions of oil and water present. In general, O/W emulsions are produced by emulsifying agents that are more soluble in the water than in the oil phase, whereas W/O emulsions are produced by emulsifying agents that are more soluble in the oil than in the water.

These two types of emulsions are easily distinguished:

(1) An emulsion can readily be diluted with more of the outer phase, but not as easily with the inner phase. Consequently, O/W emulsions disperse readily in water, W/O do not, but they do disperse readily in oil. This method works best on dilute emulsions.

(2) O/W emulsions have electrical conductivities similar to that of the water phase; W/O emulsions do not conduct current significantly.

(3) W/O emulsions will be colored by oil-soluble dyes, whereas O/W emulsions show the color faintly, if at all, but will be colored by water-soluble dyes.

(4) If the two phases have different refractive indices, microscopic examination of the droplets will determine their nature. A droplet, on focusing upward, will appear brighter if its refractive index is greater than the continuous phase and darker if its refractive index is less than that of the continuous phase. This clearly identifies the substance in the droplet, if one knows the relative refractive indices of the two phases.

(5) In filter paper tests, a drop of an O/W emulsion produces an immediate wide moist area; a drop of W/O emulsion does not. If the filter paper is first impregnated with a 20% cobaltous chloride solution and dried before the test, the area around the drop immediately turns pink if the emulsion is O/W and remains blue (shows no color change) if it is W/O.

Emulsions are significantly stable. The significantly stable means relative to the intended use and may range from a few minutes to a few years.

The rate at which the droplets of a macroemulsion coalesce to form larger droplets and eventually break the emulsion has been found to depend on a number of factors: the physical nature of the interfacial film, the existence of an electrical or stone barrier on the droplets; the viscosity of the continuous phase; the size distribution of the droplets; the phase volume ratio; the temperature.

2.3.4 Physical Nature of the Interfacial Film

The droplets of dispersed liquid in an emulsion are in constant motion and therefore there are frequent collisions between them. If, on collision, the interfacial film surrounding the two colliding droplets in a macroemulsion ruptures, the two droplets will coalesce to form a large one, since this results in a decrease in the free energy of the system. If this process continues, the dispersed phase will separate from the emulsion, and it will break. The mechanical strength of the interfacial film is therefore one of the prime factors determining macroemulsion stability. For maximum mechanical stability, the interfacial film resulting from the adsorbed surfactants should be condensed, with strong lateral intermolecular forces, and should exhibit high film elasticity.

Since highly purified surfactants generally produce interfacial films that are not close packed and hence not mechanically strong, good emulsifying agents are usually a mixture of two or more surfactants rather than an individual surfactant. A commonly used combination consists of a water-soluble surfactant and an oil-soluble one.

2.3.5 Existence of an Electrical or Steric Barrier to Coalescence an the Dispersed Droplets

The presence of a charge on the dispersed droplets constitutes an electrical barrier to the close approach of two particles to each other. This is believed to be a significant factor only in O/W emulsions. In O/W emulsions, the source of the charge on the dispersed droplets is the adsorbed layer of surfactant with its hydrophilic end oriented toward the water phase. In emulsions stabilized by ionic surfactants, the charge on the dispersed phase may arise either from adsorption of ions from the aqueous phase or from frictional contact between droplets and the aqueous phase. In the latter case, the phase with the higher dielectric constant is charged positively. In W/O emulsions, there is very little charge, if any, on the dispersed particles and experimental data indicate no correlation between stability and any charge present. In fact, for water-in-benzene emulsions stabilized by oleate soaps of polyvalent metals, an anticorrelation was found between zeta potential and stability against coalescence. The true stabilizers in these systems are probably insoluble basic metal oleates produced by hydrolysis of the original metal oleates. Those metal oleates that do not stabilize water-in-benzene emulsions show on hydrolysis and have the highest zeta potentials. The hydrolysis products, if insoluble in both phases, accumulate at the interface and prevent the formation of an electrical double layer to the oil phase. Their accumulation at the interface stabilizes the W/O emulsion, since these basic metal oleates are preferentially wetted by the benzene and, in addition, form an interfacial film or layer that machnically prevents coalescence of the water droplets.

2.3.6 Viscosity of the Continuous Phase

An increase in the viscosity η of the continuous phase reduces the diffusion coefficient D of the droplets, since, for spherical droplets,

$$D = \frac{kT}{6\pi\eta a} \tag{2-6}$$

Where k—the Boltzmann constant;
   T—the thermodynamic temperature;
   a—the radius of the droplets.

As the diffusion constant is reduced, the frequency of collision of the droplets and their rate of coalescence are reduced. The viscosity of the external phase is increased as the number of suspended particles increases, and this is one of the reasons that many emulsions are more stable in concentrated form than when diluted. The viscosity of the external phase in emulsions is often increased by the addition of special ingredients for this purpose, such as natural and synthetic "thickening" agents.

2.3.7 Size Distribution of Droplets

A factor influencing the rate of coalescence of the droplets is the size distribution. The smaller the range of sizes, the more stable the emulsion. Since large particles have less interfacial surface per unit volume than smaller droplets, in macroemulsions they are thermodynamically more stable than the smaller droplets and tend to grow at the expense of the smaller ones. If this process continues, the emulsion eventually "breaks". An emulsion with a fairly uniform size distribution is therefore more stable than one with the same average particle size having a wider distribution of sizes.

2.3.8 Phase Volume Ratio

As the volume of the dispersed phase in a macroemulsion increases, the interfacial film expands further and further to surround the droplets of dispersed material, and the basic instability of the system increases. As the volume of the dispersed phase increase beyond that of the continuous phase, the type of emulsion (O/W) or (W/O) becomes basically more and more unstable relative to the other type of emulsion, since the area of the interface that is now enclosing the dispersed phase is larger than that which would be needed to enclose the continuous phase. It often happens, therefore, that the emulsion inverts as more and more of the dispersed phase is added, unless the emulsifying agent is so unbalanced as to be capable of forming only one type of emulsion.

2.3.9 Temperature

A change to temperature causes changes to the interfacial tension between the two phases, to the nature and viscosity of the interfacial film, in the relative solubility of the emulsifying agent in the two phases, in the vapor pressures and viscosities of the liquid phases. And in the thermal agitation of the dispersed particles. Therefore, temperature changes usually cause considerable changes to the stability of emulsions; they may invert the emulsion or cause it to break. Emulsifying agents are usually most effective when near the point of minimum solubility in the solvent in which they are dissolved since at that point they are most surface-active. Since the solubility of the emulsifying agent usually changes with temperature change, stability of the emulsion usually also changes because of this. Finally, anything that disturbs the interface decreases its stability, and the increased vapor pressure resulting from an increase in temperature causes an increased flow of molecules through the interface, with a resulting decrease in stability.

New Words and Expressions

interfacial [ˌɪntə(ː)'feɪʃəl] adj. 界面的	aqueous ['eɪkwɪəs] adj. (含,多)水的
contract [kən'trækt] v. 收缩	hydrophobic [ˌhaɪdrə'fəʊbɪk] adj. 疏水的
spontaneously [spɒn'teɪnɪəslɪ] adv. 自发地	hydrophilic [ˌhaɪdrə'fɪlɪk] adj. 亲水的
spherical [s'ferɪk(ə)l] adj. 球形的	dehydration [ˌdɪhaɪ'dreɪʃən] n. 脱水(作用)
tangential [tæn'dʒenʃ(ə)l] adj. 切线的	siloxane [sɪ'lɒkseɪn] n. 聚硅氧烷

isothermally [ˌaɪsəʊ'θɜːməli] adv. 等温地	heptane ['heptein] n. 正庚烷
reversibly [rɪ'vɜːsəbli] adv. 可逆地	cationic [ˌkætɪ'ɒnɪk] adj. 阳离子的
immiscible [ɪ'mɪsəbl] adj. 不相溶的	zwitterionic [ˌzwɪtəraɪ'ɒnɪk] adj. 两性的
static ['stætɪk] adj. 静态的	amphoteric [ˌæmfə'terɪk] adj. 两性的
quiescent [kwaɪ'esənt] adj. 静止的	nonionic [ˌnɒnaɪ'ɒnɪk] adj. 非离子的
surfactant [sə'fæktənt] n./adj. 表面活性剂(的)	electrostatic [ɪˌlektrəʊ'stætɪk] adj. 静电的
lyophobic [ˌlaɪə'fəʊbɪk] adj. 疏(憎)液的	ca. = circa 大约
lyophilic [ˌlaɪə'fɪlɪk] adj. 亲液的	surface tension 表面张力
amphipathic [ˌæmfɪ'pæθɪk] adj. 两亲的	interfacial tension 界面张力
distortion [dɪs'tɔːʃən] n. 变形,扭曲	net inward pull 净向内拉为
at right angles to(with) 与……垂直	electrostatic attraction 静电引力
vapour phase 气相	lignin derivatives 木质素衍生物
surface-active agent 表面活性剂	van der Waals forces of attraction 范德华力,分子间作用力
polypropylene glycol 聚丙二醇	polyoxyethylene alkylphenol 聚氧乙烯烷基酚
emulsification [ɪˌmʌlsɪfɪ'keɪʃən] n. 乳化,乳化作用	agitation [ˌædʒɪ'teɪʃn] n. 搅拌
emulsion [ɪ'mʌlʃən] n. 乳液	macroemulsion 粗(滴)乳液
versatile ['vɜːsətaɪl] adj. 通用的,万能的	microemulsion 微(滴)乳液
suspension [səs'penʃən] n. 悬浮液	discontinuous (inner) phase 不连续相(内相)
impregnate ['ɪmpregneɪt] v. 浸渍,浸润	continuous (outer) phase 连续相(外相)
cobaltous [kəʊ'bɔːltəs] adj. 亚钴的,二价钴的	electrical conductivity 电导率
coalescence [ˌkəʊə'lesns] n. 凝结	refractive index 折射率
pack [pæk] v. 压紧	lauryl alcohol 月桂醇
lateral ['lætərəl] adj. 侧面的	frictional contact 摩擦接触
dielectric [ˌdaɪɪ'lektrɪk] n. 电介质;adj. 非传导性的	dielectric constant 介电常数
oleate ['əʊlɪeɪt] n. 油酸盐	diffusion coefficient 扩散系数
viscosity [vɪ'skɒsəti] n. 黏度,黏性	Boltzmann constant 玻耳兹曼常数
diffusion [dɪ'fjuːʒən] n. 扩散	thermal agitation 热扰(骚)动
invert [ɪn'vɜːt] v. 转相	

Part 3

Professional Literature

Unit 1　Dyeing and Finishing

Lesson 1　Dyes

We see that a textile possesses color(s). It is the dyes that impart colors to them. Before dyeing, we are supposed to select dyes that can interact well with the fibers so that we can get good property of the textile. Therefore, it is important to know the classifications of dyes.

Dyes may be classified according to their chemical structures or their methods of application. Classification of dyes according to their chemical structures is most useful to the dye chemists who may be interested in dye synthesis and the relationship between the chemical structures and properties of the dyes. Classification according to method of application is most useful to the technologists who concern themselves with coloration of textile products. The names given to the dye classes generally relate to the methods of application. All currently used dyes are listed by class in the Color Index, a set of volumes published jointly by the society of dyers and colorists in the United Kingdom and the American Association of Textile Chemists and Colorists (AATCC) in the United States.

In the following discussion of dye classes, dyes are grouped according to the fiber types for which they are most widely (though not exclusively) used.

3.1.1　Dyes Mostly Used On Cellulosic Fibers

3.1.1.1　Direct Dyes

Direct dyes have natural affinity for cellulose, so they are called by this name. Direct dyes are water soluble dyes which are used primarily for cotton and rayon, even some polyamide and protein fibers are also dyed by these compounds. These dyes have the advantage of being applied directly in a hot aqueous dye solution in the presence of common salts required to stabilize the rate of dyeing, making the process relatively inexpensive. Direct dyes are simplicity of application, but they also possess the disadvantage of poor fastness to washing, because they are primarily physically bound to the fibers and are soluble in laundering solutions.

3.1.1.2 Reactive Dyes

Reactive dyes, which were developed in the 1950s, are relatively new. They are sometimes called "the fabric reactive dyes". As the name implies, the reactive dyes chemically react with the fiber forming covalent bonds. Since the covalent bonds between dye and fiber are strong, reactive dyes have excellent wash fastness. Reactive dyes are applied most commonly to the cellulosic fibers, e.g. cotton, rayon, and linen. They are easy to apply and can produce a wide range of bright colors. Reactive dyes are a good choice for washable wool and silk because of their colorfastness to laundering.

3.1.1.3 Vat Dyes

Vat dyes are insoluble in water. By chemical reduction they are converted to soluble leuco compounds which are applied to the fibers and then are reoxidized to the insoluble form. This creates a color that has good fastness to both light and washing. Because vat dyes must be applied in an alkaline solution, they are not suitable for use with protein fibers. Primarily used to cotton and rayon, vat dyes can also act as disperse dyes on polyester. These dyes provide a wide range of colors except for reds and oranges. The main disadvantage of vat dyes is their relatively high cost.

3.1.1.4 Sulfur Dyes

Sulfur dyes are complex organic compounds that need to be converted to a water soluble form having affinity for cellulose by treatment with a reducing agent under alkaline conditions, and after application to the fiber, they must be oxidized back to their pigment form. Sulfur dyes produce mostly inexpensive dark colors, such as black, brown or navy. Colorfastness of sulfur dyes is good to washing and fair to light.

3.1.1.5 Azoic Dyes

Azoic dyes are pigments that are synthesized inside the fiber. The reaction that produces dyeing takes place when two components, an aromatic diazonium salt and an aromatic hydroxyl compound, join together to form a highly colored, insoluble compound on the fiber. These dyes are also known as ice dyes because the reaction takes place at lower temperature. They have good colorfastness to washing, bleaching, alkalis, and light. Azoic dyes provide an economical way to obtain certain shades, especially red, but they sometimes possess poor fastness to crocking.

3.1.2 Dyes Used Primarily for Protein Fibers

3.1.2.1 Acid Dyes

Acid dyes have one to four negatively charged functional groups. Applied in acid solution, they react with basic groups in the fibers structure to form ionic bonds. Because wool has both acid and basic groups in its structure, acid dyes can be used successfully on wools. These dyes are also utilized for dyeing nylon, acrylic, some modified polyester, polypropylene and ammine. They have little affinity for cellulosic fibers. Colorfastness of acid dyes varies a good deal, depending on the color and the fiber to which the dye has been applied.

3.1.2.2 Chrome or Mordant Dyes

Used on the same general group of fibers as acid dyes, chrome dyes (also called mordant dyes) are applied with a metallic salt that reacts with the dye molecule to form a relatively insoluble dyestuff with improved wetfastness and lightfastness. As the name of the dye indicates, chromium salts are most often used for the process. Especially effective for dyeing wool and silk, these dyes have excellent colorfastness, but residual metals may be harmful, especially to silk fabrics. Chrome dyes usually produce duller colors.

3.1.3 Dyes Used Primarily for Manufactured Fibers

3.1.3.1 Basic Dyes

Basic dyes contain a positive charge in their molecules that are attracted to negative groups in fibers; for this reason, they are also known as cationic dyes. Acrylic fibers usually have sulfonic acid groups added to the polymer for dyeing with basic dyes. Cationic dyeable polyester, cationic dyeable nylon, wool and silk have negative acid groups can be dyed with basic dyes. The colors that these dyes produce are exceptionally bright. But their fastness to light is not satisfactory, especially on protein fibers.

3.1.3.2 Disperse Dyes

Disperse dyes are used mostly for polyester, nylon, and cellulose acetate. Meanwhile they can be used for other fibers. The name of disperse dye comes from the fact that these dyes are sparingly soluble in water and have to be dispersed in water to make the dyebath. Disperse dyes were developed when cellulose acetate was first marketed. Disperse dyes are the only acceptable dye class for acetate and unmodified polyester fibers.

3.1.4 Dye Classes According to Chemical Structure

Dye classes according to chemical structure are azo dyes, anthraquinone dyes, benzoquinoline dyes, polycyclic aromatic carbonyl dyes, polymethine dyes, styryl dyes, aryl carbonium dyes, phthalocyanine dyes, benzofuran dyes, sulfur dyes, nitro and nitroso dyes. Chemical classification is primarily made with regard to functional groups and/or ring systems, with the lowest numbers (10,000 to 10,299) assigned to nitroso or quinone oxime structures and the highest numbers (77,000 to 77,999) assigned to inorganic coloring matters (usually pigments). The highest numbers in the chemical classification scheme generally reflected more complex heterocyclic ring structure, while the lower numbers in this chemical classification are typically carbocyclic aromatic rings containing various types of functional groups.

New Words and Expressions

fastness ['fɑːstnɪs] n. 牢度,不褪色(性)	acrylic [ə'krɪlɪk] n. 聚丙烯腈系纤维
laundering ['lɔːndərɪŋ] n. 洗烫	polypropylene [ˌpɒlɪ'prəʊpɪliːn] n. 聚丙烯
dyestuff ['daɪstʌf] n. 染料,着色剂	aramine ['æmɪːn] n. 氨络物
leuco ['ljuːkəʊ] n. 隐色体; adj. 无色的	wetfastness ['wetfɑːstnɪs] n. 湿牢度
reoxidize ['riːɒksɪdaɪz] v. 再氧化	lightfastness [laɪtfɑːstnɪs] n. 耐光牢度
pigment ['pɪgmənt] n. 颜料	sulfonic [sʌl'fɒnɪk] adj. 磺酸基的
azo ['æzəʊ] n. 偶氮,偶氮基	basic dyes 碱性染料
azoic [ə'zəʊɪk] adj. 不溶性偶氮的	cationic dyes 阳离子染料
nitro ['naɪtrəʊ] n. 硝基	cellulose acetate 醋酸纤维素
nitroso [naɪ'trəsəʊ] n. 亚硝基	anthraquinone dyes 蒽醌染料
covalent bond 共价键	polymethine dyes 聚甲炔染料
vat dyes 还原染料	styryl dyes 苯乙烯染料
azoic dyes 不溶性偶氮染料	phthalocyanine dyes 酞菁染料
chrome dyes 铬媒染料	aryl carbonium dyes 芳基甲烷染料
mordant dyes 媒染染料	benzoquinoline dyes 苯并喹啉染料
chromium salts 铬盐	benzofuran dyes 苯并呋喃染料

Lesson 2 Fibers

3.2.1 Natural Fibers

Fiber, the primary material from which most textile products are made, can be defined as units of hairlike dimensions, with a length at least one hundred times greater than the width. The textile fibers may be divided into two major groups according to their origin: ①natural fibers, ②chemical fibers.

Those that are found in nature are known as natural fibers, which are taken from either animal, vegetable or mineral resources. The classification of natural fibers is shown in Table 3-1. Vegetable fibers could be further divided according to the part of the plant that produces the fiber: the leaf, a hair produced from a seedpod or the stem. The latter is called bast fibers. Animal fibers could be further divided into those fibers from the hair of an animal such as wool and those from an extruded web such as silk. Chemically, the classification might be cellulosic for vegetable fibers, protein for animal fibers, and name of the specific minerals (such as asbestos) for mineral fibers. Using this scheme, cotton is a plant seedpod, or cellulosic fiber and wool is an animal, hair, or protein fiber. They vary considerably as regards their properties and their productions.

Table 3-1 Classification of natural fibers

Cellulosic fibers	Protein fibers	Mineral fibers
Cotton	Wool	Asbestos
Flax	Silk	
Jute	Mohair	
Ramie	Cashmere	
	Other animal hair	

3.2.1.1 Cotton

Cotton is by far the most important textile fiber and makes up nearly 50 percent of the total mass of fibers used in the world. It is obtained from the cotton plant which grows in warm moist climates and in most parts of the world. Cotton fibers are composed largely of cellulose. Besides cellulose, raw cotton contains a number of other substances, notably waxes, pectic products and mineral substances. They can range from 4% to 12% together and are referred to as impurities by the manufacturer of cotton goods. Generally these are objectionable effects and would make it difficult to color and finish cotton fabrics satisfactorily, so it is always a first step in the art of dyeing and finishing to purify the cotton as completely as possible.

Celluloseis considered to be a condensation polymer formed from the glucose units. There are 3,000-5,000 glucose units joined together in natural cellulosic fibers. This corresponds to a molecular mass of 300,000-500,000.

Cotton is excellent for a multitude of purposes and has virtually universal consumer acceptance. It is used for apparel fabrics, for household or domestic goods, and for industrial applications. Cotton is also extensively used in blends with man-made fibers to achieve new combinations of properties that are not available in the fibers separately.

Cotton has some disadvantages, too. It creases and wrinkles easily. It is readily attacked by acid reagents or substances, and it is slowly affected by sunlight, causing yellowing and fiber degradation (see Table 3-2).

Table 3-2 Properties of main natural fibers

Property			Evalution	
Fiber	Cotton	Flax	Wool	Silk
Luster	Low	High	Medium-medium high	High
Strength	3.0-6.0 grams per denier, medium strength	2.6-7.7g/D	Dry:1.0-1.7g/D Wet:0.8-1.6g/D	Dry:2.4-5.1g/D Wet:2.0-4.3g/D
Elastic recovery/ Elongation	Low elasticity, 75% recovery at 2% extension; 3%-10% elongation	65% recovery at 2% extension; 2.7%-3.3%	99% recovery at 2% extension; extends 20%-40% when dry and 20%-70% or more when wet	92% recovery at 2% extension; Elongation: dry:10%-25% wet:33%-35%
Resiliency	Low	Poor	Excellent	Medium
Relative density	1.54-1.56	1.50	1.30-1.32	1.25-1.34
Moisture absorption	8.5% at standard conditions (sc)	12% at sc	13.6%-16% at sc	11.0% at sc
Dimensional stability	Considered relatively stable	Good	Subject to felting and relaxation shrinkage	Good
Resistant to:				
acids	Damage, weaken fiber	Good to cool, dilute acids; low or poor to hot, dilute; poor to concentrated either hot or cool	Good resistance to dilute acids; medium to poor resistance to concentrated acids	Low, dissolves or damaged by most mineral acids; organic acids do not damage
alkalis	Resistant, no harmful effects	High resistance	Low resistance; many alkalis destory fiber	Strong alkalis damage fiber; weak have litter or no effect
sunlight	Prolonged exposure weakens fibers	Good	Prolonged exposure deteriorates fibers	Prolonged exposure causes fibers breakdown
microorganisms	Mildew damages fibers, rot-bacteria damages the fiber	Mildew will grow on and damage fibers, particularly in humid atmosphere	Resistance generally good	Good resistance
insects	Silverfish damage cotton	Good	Damage by insects such as moths and carpet beetles	Destoryed by carpet bettles
Thermal reactions				
to heat	Decompose following extended subjection to heat at 150℃	Gradual decomposition after prolonged exposure at 150℃	Avoid prolonged exposure to temperatures above 130℃	temperatures above 150℃ result in yellowing and general discoloration
to flame	Burn readily	Burn readily	Burn slowly when in direct flame; considered to be self-extinguishing	Burn with a sputtering flame

3.2.1.2 Flax (Linen)

Flax is a bast fiber obtained from the flax plant. Flax fibers resemble cotton in so far as they consist of cellulose but here lower cellulose content. On an average the flax fibers contain only about 75 percent of pure cellulose, the remaining matter being a gummy pectic substance. The polymer of flax consists with a degree of polymerization of about 36,000 glucose units.

The surface of each fiber is smooth and this helps to give linen materials their characteristic high luster. Flax has relatively high strength. In many of its chemical properties linen closely resembles cotton. Thus, it is resistant to alkalis and is easily deteriorated by acids. Properties of flax fibers are shown in Table 3-2. Linen is mainly used in the manufacture of sail cloth, tent fabric, sewing threads, fishing lines, table-cloth and sheets, but today it is often used as a component of blends.

3.2.1.3 Wool

Wool is the most important fiber and produced in the largest amount. Wool is the fur-like covering of sheep that are raised in many countries arround the world. It is obtained by shearing the fibrous covering from the sheep.

Wool fibers are composed of protein in which the repeated unit is amino acid. The amino acids are linked to each other by the peptide bond (—CO—NH—) to form the protein polymer. Chemically, the most important component in wool is keratin that is a complex protein and composed of 16 to 18 different amino acids. Kertain is amphoteric in nature. So wool can be dyed with acid or reactive dyes.

Most wool fibershave a white or creamy color, although some breeds of sheep yield brown or black wool. Wool fibers have a tendency to return completely to their original shape after small deformations, which is great asset in apparel fabrics. The natural crimp in wool is of great importance, since it results in making a yarn fluffy, thereby trapping air in the interstices between the fibers. This trapping of air helps in forming an insulating layer, thus imparting the characteristic of warmth. Wool has several disadvantages: it is very sensitive to alkaline substance, it is readily attacked by moths and carpet beetles unless treated to resist them, it is difficult to bleach, and it felts easily.

According to their finenessand length wool can be divided into four types: fine wool, long wool, medium wool and carper wool. Wool is used primarily in apparel and home furnishings.

3.2.1.4 Silk

Silk is the material extruded from gland in the body of the silkworm in spinning its cocoon or web. To reclaim silk filaments, the cocoons are soaked in hot water, which soften the sericin gum. Filament from several cocoons are picked up, assembled, passed through a guide, and made into skeins by the process of reeling. This yarn can then be processed into fabrics before or after degumming.

Raw silk is composed of two proteins: fibroin and sericin. Fibroin is the actual fiber protein. Sericin is the gummy substance and holds the filament together. The average composition of raw silk is 70%-75% fibroin, 20%-25% sericin, 2%-3% waxy substances and 1%-1.7% mineral matter. Both of fibroin and sericin are built up of 16-18 amino acids. The degree of polymerization of silk fibroin is uncertain, with DP of 300 to 3,000 having been measured in different solvents.

Silk fibers have smooth surface and a distinctive triangular cross-sectional dimension, so silk fibers have a lustrous appearance. Silk has high tenacity and relatively good moisture regain. Silk is warm and pleasant to touch and is the generally considered comfortable to wear. It is readily dyeable with a variety of dyes. Silk is popular in men's neckties for its hand and drape. The fiber is used alone and in blends with other fibers.

3.2.2 Chemical Fibers

Those that are created through chemical technology are known as chemical fiber. Most of chemical fibers are formed by forcing a viscous chemical substance through a spinneret which consists of a series of tiny holes arranged in a circle. The streams emerging from the holes are then hardened or solidified to form filaments. Chemical fibers are subdivided into two basic classifications: ①regenerated fibers and ②synthetic fibers. The regenerated fibers are those in which the fiber-forming material is of natural origin; the second class of fibers is made by the chemical synthesis of simple polymer-forming materials. The classification of chemical fibers is shown in Table 3-3.

Table 3-3 Classification of chemical fibers

Regenerated fibers	Synthetic fibers
Viscose rayon	Polyamide
Cuprammonium rayon	Polyester
Diacetate	Polyacrylonitrile
Triacetate	Elastane
Protein regenerated fiber	Polyolefin
Tencel	Polyvinyl

3.2.2.1 Regenerated Fibers

Thereare five types of regenerated fibers: viscose rayon, cuprammonium rayon, tencel, acerate fiber and protein regenerated fiber. The first four are manufactured from a natural polymer (cellulose), which is usually obtained from wood and cotton linters. The latter may be produced from animal and vegetable proteins.

(1) Viscose Rayon

Viscose rayon is produced in common by the so-called wet-spinning process. The process involves the reaction of cellulose with sodium hydroxide to form alkali cellulose. The alkali cellulose is then treated with carbon disulfide to form cellulose xanthate. This material is dissolved in dilute sodium hydroxide to form the viscose solution. Viscose is then extruded through special orifices into an acid spinning bath in which the filament is coagulated and sodium hydroxide is neutralized. Finally, the filament is collected and combined into yarns as end product.

Viscose rayon is used in a multiplicity of textile applications. It is comparatively low in cost. High-tenacity rayons are used in tire cords and various industrial applications. Rayon staple is used extensively in blend with cotton, wool or any other man-made textile fibers. Rayon is absorbent and, therefore, comfortable to wear. However, its greatest drawbacks are its loss of strength on wetting and its dimensional instability.

(2) Cuprammonium Rayon

Regenerated cellulose fibers produced from the solution of cellulose in a mixture of copper sulfate and ammonia is called cuprammonium rayon or "Cup".

Cuprammonium rayon filament is produced only on a relatively small scale as the manufacture of the fiber gives rise to a serious environmental problem. Many other physical characteristics are similar to those of viscose. Due to high amorphous structure the filaments can absorb more water than cotton. The cupre filaments have a silk like appearance. So it is used for the manufacture of the fabrics and garments and lady's hose where it is desired to imitate silk in handle and appearance.

(3) Acetate Fibers

Acetate fibers are also produced from cotton inner, or purified wood pulp. In this group there are two fibers: diacetate and triacetate fibers.

The production of diacetate fiber includes the following stages:

The natural cellulose is firstly acetylated and converted into cellulose triacetate at temperature up to 50℃ with acetic anhydride in the presence of glacial acetic acid and concentrated sulphuric acid.

And hydrolyzed to form cellulose diacetate through an ageing or ripening process in presence of water.

The resulted cellulose diacetate flake is then dissolved in acetone containing 4% water as the solvent to form the spinning dope, which is filtered and then force through the spinneret into a warm-air chamber and the method of spinning is called dry-spin.

Triacetate is manufactured from the same raw materials as cellulose diacetate, bur the ripening stage in which hydrolysis occurs is omitted in triacetate production. To produce spinning solution, dried acetylated flake is dissolved in methylene chloride and dry-spun into a warm-air chamber.

Both of them are thermoplastic fiber and have certain recovery. They can be dyed with disperse dyes. Diacetate fiber is softer than triacetate and used mainly for apparel fabrics such as satins and taffetas, most cigarette filter tips are made from the cellulose diacetate fibers. While triacetate fiber is used in a multiplicity of textile application, such as knitted and woven undergarments, skirts and slacks materials, table clothes, furnishing fabrics, etc.

(4) Tencel

Tencel, which is called Lyocell, is a regenerated cellulose fiber. It is produced from purified conifer wood pulp through solvent spinning process, in which the purified cellulose is dissolved in the solvent of N-methyl amine oxide to form spinning dope, the spinning dope is filtered, and then the extruded through a spinnerette to form a filament.

The solvent used in the manufacture oftencel is atoxic and more than 99.5% of it can be reclaimed and reused. So the manufacture of the fiber does not give rise to any environment problem and is known as "green process".

Tencel fiber can be degraded by the action of microbe, so it is also considered a "green fiber". Tencel fiber has high tensile strength and high moisture absorption capacity. It can provide good wearing performances. It blends well with cotton, wool, silk and other fibers. Fabrics made from it or its blends possess a soft hand, pearl-like luster and good drape. It is primarily used for lady's underwear, shirts, slacks, etc. which are associated with fashion.

(5) Regenerated Protein Fiber

The raw materials used for the preparation of regenerated protein fiber may be milk, soybeans, peanuts and zein. Sometimes alkaline solutions of gelatin, albumin and other raw materials like waste wool, silk and feathers may be used.

Soybean protein fiber is the new type textile material, which uses the advanced technology to develop. It has been praised locally and internationally by industry expert as the "Healthy and Comfortable Fiber of the 21st Century".

Properties of this fiber are excellent. It has many merits of natural fiber and chemical fiber: thinness, lightness, high strength (breaking tenacity), good resistance of acid and alkali, excellent wet-absorption and wet-transference etc.

Pure soybean yarn or soybean blend with cotton yarn can be used to produce knitting fabric or weaving fabric which are suit for underskirt and garment.

3.2.2.2 Synthetic Fibers

The first synthetic fiber is nylon (one of polyamide fibers), which was commercially produced in United States in 1939. The main kinds of the synthetic fibers include: polyamide, polyester, polyactylonitrile fibers and spandex fibers, which are used in textile industry widely.

(1) Polyamide Fibers

Among the polyamide fibers, nylon 66 end nylon 6 are used more frequently. Nylon 66 is produced by polycondensation from adipic acid and hexamethylene diamine. Nylon 6 is the polymerization product made from caprolactam. Both of them are manufactured through the fiber production process called melts pun in which the polymer is converted to a liquid form just by heat and then extruded through a spinneret to form filament.

Nylon is avery strong, quick-drying fiber with high wet strength and has excellent elasticity. Nylon has a lower specific gravity than other fibers. These properties make them very suitable for stockings, parachute fabrics, shirts, underwear, carpet and reinforcement of rubber in tyres and belts. Nylon blends well with other fiber and adds strength to such blends (15 to 20 percent nylon is needed to give additional strength to most fabrics).

(2) Polyester Fibers

Polyester fiber is now the largest man-made fiber in terms of volume of production. Polyester polymer that is often manufactured in chemical factories is obtained by the condensation of terephthalic acid or dimethyl terephthalate with ethylene glycol at high temperature.

The outstanding characteristics of polyester are wrinkle-free appearance and case of care. The fabrics require little or no ironing; they are easy to launder and quick to dry. The polyesters are very strong fibers; therefore, strong fabrics can be made from them. Polyesters blend well with other fibers, contributing easy maintenance, strength and durability, abrasion resistance, relatively wrinkle-free appearance, and shape and size retention. In blended fabrics, protein or cellulosic fibers enhance dyeability, comfort, absorbency, reduce static charges.

(3) Polyacrylonitrile Fibers

Polyacrylonitrile fibers include acrylic and modified acrylic fibers. All of them are made from the copolymers of acrylonitrile with comonomers such as acrylic acid, vinylidine chloride, etc.

The polymer of acrylonitrile has some undesired properties, so the comonomers are added to increase the polymer thermoplasticity, solubility, dyeability, moisture regain, etc. Acrylic fibers are composed of at least 85% by mass of acrylonitrile and 15% or less comonomer; but the modified acrylic fibers such as modacrylic fiber are comprised of less than 85 % but at least 35% by mass of acrylonitrile.

Acrylic fibers are relatively strong. They have good elasticity and high bulking power with excellent resistance to sun and weather. Blends of acrylic fiber with wool, cotton, rayon, and nylon, are common. Acrylic fibers dye easily, offering a large selection of colors. Acrylic fibers are found many applications in knitwear, carpets and pile fabrics.

(4) Spandex Fibers

Spandex is defined by TEPIA as follows: a manufactured fiber in which the fiber-forming substance is a long-chain synthetic polymer comprised of at least 85 percent by weight segmented polyurethane.

DuPont introduced a new stretch fiber in 1958. The name Lycra was given to the fiber in late 1959, and plans were announced for volume production. The name Lycra is still used of DuPont's fiber.

Spandex fibers are generally stronger than ordinary rubber filaments. The tenacity of these fibers varies in the range of 0.55-1.0g/D as compared to 0.25g/D for natural rubber. Their breaking elongation may vary up to 700%, and they demonstrate excellent recovery behavior. They have very low moisture regain. Spandex fibers have a good resistance to acids, alkalis and most common chemicals.

Spandex fibers are used in swimwear, foundation garments, support hosiery, sack tops, elastic webbing for waistbands and other similar uses. There is a great deal of interest in the use of spandex inactive-sportswear fabric.

New Words and Expressions

textile ['tekstaɪl] adj. 纺织的；n. 纺织品	fabric ['fæbrɪk] n. 织物
cellulosic [seljʊ'ləʊsɪk] adj. 纤维素的	glucose ['gluːkəʊs] n. 葡萄糖
asbestos [æs'bestəs] n. 石棉	blend [blend] n./v. 混纺
flax [flæks] n. 亚麻	degradation [,degrə'deɪʃn] n. 降解
jute [dʒuːt] n. 黄麻	linen ['lɪnɪn] n. 亚麻
ramie ['ræmɪ] n. 苎麻	elongation [,iːlɒŋ'geɪʃn] n. 伸长
mohair ['məʊheə(r)] n. 马海毛,安哥拉山羊毛	resiliency [rɪ'zɪlɪənsɪ] n. 弹性
cashmere ['kæʃmɪə(r)] n. 开司米,山羊绒	denier ['denɪə(r)] n. 旦尼尔
pectic ['pektɪk] adj. 果胶的	extension [ɪk'stenʃn] n. 伸长
keratin ['kerətɪn] n. 角蛋白	reel [riːl] v. 缫丝,络丝
deformation [,diːfɔː'meɪʃn] n. 变形	degumming [diː'gʌmɪŋ] v. 脱胶
fluffy ['flʌfɪ] adj. 绒毛似的,蓬松的	hand [hænd] n. 手感
bleach [bliːtʃ] v. 漂白	bast fiber 韧皮纤维
felt [felt] v. 缩绒	naturalfiber 天然纤维
apparel [ə'pærəl] n. 服装	chemical fiber 化学纤维
gland [glænd] n. 腺体	mineral fiber 矿物纤维
spin [spɪn] v. 纺纱	moisture absorption 吸湿性
reclaim [rɪ'kleɪm] v. 缫丝,回收	cross-section 横截面
filament ['fɪləmənt] n. 长丝	moisture regain 回潮率
soak [səʊk] v. 浸,泡,浸透	dimensional stability 尺寸稳定性
sericin ['seərɪsɪn] n. 丝胶	yellowing 泛黄
guide [gaɪd] n. 导丝器	peptide bond 肽键
skein [skeɪn] n. 绞纱	acid dyes 酸性染料
tenacity [tə'næsətɪ] n. 韧性,强度	reactive dyes 活性染料
luster ['lʌstə] n. 光彩,光泽	spinneret ['spɪnəret] n. 喷丝头
viscose rayon ['vɪskəʊs]['reɪɒn] n. 黏胶纤维,人造丝	diacetate fiber 二醋酸纤维
polyolefin [,pɒlɪ'əʊləfɪn] n. 聚烯烃	triacetate fiber 三醋酸纤维
polyvinyl [,pɒlɪ'vaɪnɪl] n. 聚乙烯	Tencel fiber 天丝纤维(商品名)
linter ['lɪntə] n. 棉短绒	polyamide fiber 聚酰胺纤维,锦纶
xanthate ['zænθeɪt] n. 黄原酸盐(或酯)	wet-spinning process 湿纺工艺
orifice ['ɒrɪfɪs] n. 喷丝孔,小孔	alkali cellulose 碱纤维素
coagulate [kəʊ'æɡjuleɪt] v. 凝固	acetic anhydride 醋酸酐
acetylate [æ'sɪtɪleɪt] v. 乙酰化	glacial acetic acid 冰醋酸

hydrolyze ['haɪdrəlaɪz] v. 水解	ripening process 熟化工艺
satin ['sætɪn] n. 缎,缎纹织物	spinning dope 纺丝液
taffeta ['tæfɪtə] n. 塔夫绸,平纹皱丝织品	dry-spinning 干法纺丝
knitted [nɪtɪd] adj. 针织的	disperse dye 分散染料
woven ['wəʊvn] adj. 机织的	N-methyl amine oxide N-甲基氧化胺
zein ['ziːɪn] n. 玉米蛋白	tensile strength 拉伸强度,断裂强度
gelatin ['dʒelətɪn] adj. 凝胶,明胶	soybean protein fiber 大豆蛋白纤维
albumin [æl'bjʊm] n. 清蛋白,白蛋白	breaking tenacity 断裂强度
caprolactam [ˌkæprəʊ'læktəm] n. 己内酰胺	adipic acid 己二酸
acrylic [ə'krɪlɪk] n. 聚丙烯酸; adj. 丙烯酸的	hexamethylene diamine 己二胺
Lycra ['laɪkrə] n. 莱卡(商品名)	terephthalic acid 对苯二酸
commonomer [kəʊ'mɒnəmə] n. 共聚单体	ethylene glycol 乙二醇
cuprammonium rayon 铜氨人造丝	abrasion resistance 耐磨性
rayon staple 黏胶短纤维	acrylic acid 丙烯酸
polyacrylonitrile fiber 聚丙烯腈纤维,腈纶	vinylidine chloride 偏二氯乙烯
meltspun 熔融纺丝	spandex fiber(斯潘德克斯)弹性纤维,氨纶

Lesson 3 Color

Colors are sensory perceptlons produced when light waves reflected from an object strike the eye. Color is the phenomenon which allows one to differentiate otherwise visually identical objects. Everything has color if we indude white and black as colors. The color of textile products has a profound effect on the appeal of the products to the consumer.

The existence of color requires a source of light, an object, and an observer to see the light. All three of these conditions are variables in the perception of color.

3.3.1 Light

Visible light consists of the narrow band of the electromagnetic spectrum having wavelengths in the range of about 380-780 nanometerrs (nm). While visible light is most often specified by its wavelength, it can be denoted as well by its frequency or wavenumber. Frequency is the number of wave cycles occurring per unit of time. Frequency is most often specified as cycles per second, or

Hertz (Hz), and is inversely proportional to wavelength.

$$\text{frequency (Hz)} = \text{speed of light (length/time)} / \text{wavelength (length)}$$

Wavenumber is the number of wave cycles occurring per unit of length and is the reciprocal of wavelength.

$$\text{wavenumber} = 1/\text{wavelength}$$

The approximate wavelengths of visible light that produce various color perceptions are shown in Table 3-4.

Table 3-4 Colors of the visible spectrum

Color Perceived	Wavelengths/nm
Red	700～610
Orange	610～590
Yellow	590～570
Green	570～480
Blue	480～430
Violet	430～390

Since an object can only reflect light waves which shine on the object, the color of the object depends on the intensity of the various wavelengths which are present in the light source. A light source is defined by its spectral energy distribution. Blackbody temperature is a concept based on the fact that the relative spectral energy distribution emitted by an object depends on the temperature of the object. For example, when a metal is heated it glows, first becoming reddish and then progressively whiter and brighter as its temperature rises. A blackbody is a real or theoretical object that emits a certain energy distribution depending on its temperature. Light sources are sometimes described by their "color temperature", the blackbody temperature in K which produces the spectral energy distribution approximating the light source.

Commission Internationale de L'Eclairage (CIE) defines several standard light sources. Having standard light sources is important so that standard conditions can be established for viewing, comparison, acceptance, and rejection of colors in commerce.

CIE sources A, B, and C are all incandescent. Source A is a tungsten-filament lamp operating at a color temperature of 2854K. Source A is more intense in the long wavelengths making it appear reddish. Sources B and C are filtered tungsten-filament sources operating at color temperatures of 4800K and 6500K, respectively. Source B is intended to simulate noon sunlight while source C is intended to simulate overcast-sky daylight.

The tungsten-filament sources described above approximate the appearance of daylight but contain less ultraviolet waves than natural daylight. Because many textile products and colors are fluorescent, their appearance changes depending on the ultraviolet wave content of the light source. The CIE specified a series of illuminants to supplement A, B, and C. Illuminant D_{65} (color temperature of 6500K) is now widely used in colorimetry. D_{65} and D_{75} (colors temperatures of 5500K and 7500K) were also specified by the CIE. Note that the term illuminant is used instead of source when referring to the D illuminants. A source is a real light that can be physically turned on and off. An illuminant is an imaginary spectral power distribution. It may or may not be possible to make a physical source identical to a defined illuminant.

Some light sources are continuous meaning that they contain all wavelengths of the visible spectrum. Other light sources are discontinuous having all of their energy concentrated in a few narrow bands. These are called line sources. Mercury lamps and fluorescent lamps are examples of line sources. "Prime color" fluorescent lamps have high relative energy output in the wavelength regions which stimulate most srrongly the color receptors in the human eye. These sources make colors appear especially intense and bright.

3.3.2 Object

The modification of incident light by an object affects the color perceived by the observer. Following are some ways that an object can modify light.

(1) Transmit

Light passes through without a change in direction. The object is said to be transparent. If some wavelengths are transmitted to a greater degree than others, the object is colored as opposed to white or colorless.

(2) Reflect

Light waves bounce off at an angle equal to the angle of incidence in "specular reflection". If certain wavelengths of light are reflected to a greater degree than other wavelengths, the object has color as opposed to being white. If no light is reflected by the object, the object is black.

(3) Scatter

Light waves bounce off at various angles. This is called scattering or "diffuse reflection". If all of the incident light is reflected, scattered, or absorbed, and no light is transmitted; the object is opaque. If part of the light is transmitted, the object is said to be translucent. Light scattering is an important phenomenon which is responsible for the blue color of the sky and colors of smoke and clouds. Light scattering greatly influences the color of textile materials.

(4) Absorb

Light is transformed to some other form of energy, typically heat. Much of the light not transmitted, reflected, or scattered is absorbed.

(5) Refract

Speed of light is changed as light passes through an object resulting is bending of the waves. Refraction is responsible for the appearance of a dividing line between two liquids in contact with one another and for the apparent bending of a stick at the point that it is submerged in water.

(6) Fluoresce

Light of some wavelengths is absorbed by an object and quickly emitted at longer wavelengths. Fluorescence is responsible for brightening effect of optical whiteners in textiles and for the brightness of certain dyes when viewed under a source containing ultraviolet waves.

(7) Phosphoresce

As in fluorescence, light of some wavelengths is absorbed and emitted at longer wavelengths. However, a phosphorescent object may the energy for an extended period of time before emitting it while a fluorescent object emits the absorbed light almost instantaneously.

3.3.3 Observer

We usuallythink of the observer as a person, but an electronic instrument also satisfies this requirement for color. The human observer sees color when light enters the eye and strikes the retina. The term "color blind" is often used to refer to an individual having a color vision deficiency. Although defective color vision is common, the term "color blind" is an overstatement in most cases since total in ability to see color is very rare. The most common color vision deficiency is difficulty in distinguishing between red and green, but other types of deficiencies also exist. About 8% of all men and 5‰ of women suffer color vision deficiencies ranging from very mild to severe.

People having normal color vision see wavelengths of light in the range about 380-780 nanometers (nm). The sensitivity of the eye is low near both ends of this range, so the "visible region" is often said to be 400-700nm. Nerves in the retina send signals to the brain so that the observer perceives an image. The retina contains two types of cells, rods and cones. The rods are very sensitive but only to wavelengths near 505 nm. Rods make it possible for humans to see in low light conditions, but they do not play a role in color perception. The cones are much less sensitive to light than the rods and vary in sensitiviry to different wavelengths of light. The cones are responsible for the perception of color. The maximum sensitivity of eye to wavelengths of about 450nm, 540nm, and 575nm is the basis for the tristimulus theory of color vison. This theory is often referred to as the Young-Helmholtz theory after its early proposers. The theory contends that the eye contains pigments which absorb light at or near the above wavelengths and undergo chemical reactions. The details of how messages are transmitted from the cones in the eye to the brain are not completely understood and are not within the scope of this text.

The opponent theory of color vision contends that the eye perceives one of each of three opponent pairs (red/green, blue/yellow, and black/white) but not both of a particular pair simultaneously. For example, a greater perception of red would be accompanied by a corresponding lesser stimulus of green. According to the opponent theory, the stimulus produced by each opponent pair contributes to the color perceived by the observer.

The spectralenergy distribution of a source can be defined and controlled very accurately. The modification of light by an object is inherent in the object and can be accurately measured. These elements in the perception of color can be handled very objectively. On the other hand, acuity of color vision varies greatly among individuals. Even individuals considered to have normal color vision probably see colors slightly differently, and the psychological response to color is certainly very different from one individual to another. Therefore, the perception of color by an individual is a subjective matter, a fact which can have significant influence in acceptance or rejection of colors in commerce.

3.3.4 Color Mixing

When colors are mixed, new colors result. The color perceived by an observer results from the stimulus produced by the mixture of wavelengths of light that enter eye. A color mixture which results in an increase in the number of wavelengths or intensity of light entering the eye is an "additive color mixture". Mixing of lights is an example of additive color mixing. Figure 3-1 shows the colors produced by additive mixing of red, blue, and green primary lights.

The circles in Figure 3-1 can be viewed as spots from colored lamps on a white screen. Where the spots overlap, the amount of light reflected by the screen is greater and a brighter secondary color is formed. Where the light from all three primary colors overlap, the reflected light mixture is white. Additive color mixing is used to produce colors in color television picture tubes. A color mixture which results in a decrease in the number of wavelengths or intensity of light entering the eye is a "subtractive color mixture". Subtractive color mixing results from mixing of dyes, pigments, or inks that absorb light. Figure 3-2 shows the effect of subtractive mixing the primary colors yellow, cyan and magenta. The secondary colors produced are duller because the total light reflectance is lower than that of either of the primary colors. Where all three of the primaries are mixed the color is black. Subtractive color mixing is used in dyeing of textile materials, formulation of paints, and for coloration of most objects.

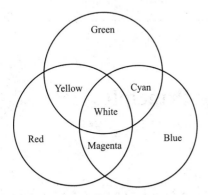
Figure 3-1 Additive color mixing

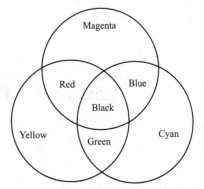
Figure 3-2 Subtractive color mixing

Spectrophotometric reflectance curves are best interpreted by thinking in terms of additive mixtures. For example, a dyer starts out with white cloth. If an orange color is needed, the dyer uses an orange dye or a mixture of a red and a yellow dye. As is indicated in Figure 3-3, the resulting fabric absorbs most of the violet, blue, and green wavelengths and reflects mostly red, orange, and yellow. The observer sees reflected light and the stimulus produced is the additive mixture of the reflected colors.

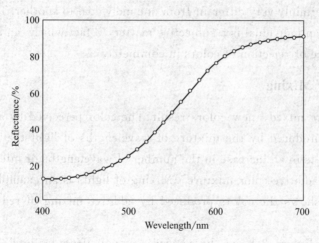

Figure 3-3 Reflectance curve of an orange color

New Words and Expressions

sensory perceptlon 感觉	incandescent [ˌɪnkæn'desnt] adj. 白炽的
strike [straɪk] n. 刺激	tungsten-filament lamp 钨丝灯
electromagnetic spectrum 电磁光谱	illuminant [ɪ'l(j)uːmɪnənt] n. 照明体
frequency ['friːkwənsɪ] n. 频率	colorimetry [ˌkʌlə'rɪmətrɪ] n. 色度学
reciprocal [rɪ'sɪprəkl] n. 倒数	mercury lamps 汞弧灯
spectral energy distribution 光谱能量分布	ultraviolet [ˌʌltrə'vaɪəlet] adj. 紫外的
blackbody ['blæk'bɒdɪ] n. 黑体	fluoresce [flʊə'res] n. 荧光
Commission Internationale De L'Eclairage (CIE) 国际照明委员会	phosphoresce [ˌfɒsfə'res] n. 磷光

Lesson 4 Dyeing Principles

Processes through which textile materialsare dyed in solution and exhibit substantivity for a dye are normally subdivided into (a) those are kinetically controlled by various diffusion and transport phenomena, and (b) those are equally controlled by sorption and desorption processes which occur between a fibrous substrate and the dye to solution. The simplest relationship between the rate of dyeing a fiber (F) in a dye solution (S) at a constant dye concentration (C_S) and temperature (T) and its equilibrium uptake of dye from that solution is shown in Figure 3-4.

The process of anchoring dyestuff molecules to the textile fibers consists of three sta-

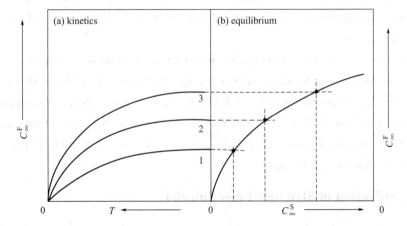

Figure 3-4 The relationship between the rate of dyeing a fiber (F) in a dye solution (S) at a constant dye concentration (C_s) and temperature (T) and its equilibrium uptake of dye from that solution

ges, which are:

(1) Migration of the dye from the solution to the interface, accompanied by adsorption on the surface of the fiber.

(2) Diffusion of the dye from the surface towards the center of the fiber.

(3) The anchoring of the dye molecules by covalent or hydrogen bonds, or other forces of a physical nature.

The assembly of the dye molecules at the surface of the fiber (stage 1) is governed, in the main, by three influences: firstly electro-potential forces, secondly temperature, and thirdly agitation.

All textile fibers, when immersed in water or aqueous solutions, acquire an electric potential, often referred to as the zeta potential.

3.4.1 Convective Diffusion Processes

The first process that occurs is convective diffusion, that is, diffusion of the dye from solution to the fiber surface. This diffusion process is influenced by such factors: the hydrodynamic flow of the dye solution, the diffusional boundary layer at the fiber surface, the zeta potential of the fibers and state of aggregation or association of dye molecules in solution.

When the dye solution initially makes contact with the fiber surface, its concentration is lower near the surface. Convective transport of the solution or aqueous diffusion of the solution supplies additional dye to the surface. Models have been devised to identify and differentiate the two boundary layers present at the fiber surface. They are the hydrodynamic boundary layer and the diffuse boundary layer. At high rates of flow, the dye solutions have decreasing rates of velocity as they approach the fiber surface. The hydrodynamic layer is approximately ten times thicker than the diffusional boundary layer. Hydrodynamic and diffusional boundary layer effects may be overcome by vigorous agitation of the fibrous substrate in the dyebath and/or increasing the velocity of the dye solution. This effectively makes it easier for the dye to reach the fiber surface and thus increases the dyeing rate.

Since the zeta potential of fibers causes electrostatic repulsion or attraction of dye molecules, it can be used to accelerate or retard dye uptake by changing the dyebath temperature and pH, or by adding electrolytes to the dyebath. An increase in dyebath temperature usually causes disaggregation or dissociation of dyes and improves their rate of diffusion to the fiber surface. Generally speaking, hydrodynamic and diffusional boundary layers in steady laminar flow parallel to a plane slab.

3.4.2 Diffusion of Dye into Fiber Interior

Diffusion of the dye from the fiber surface to the interior is kinetically controlled. Because the rate of dyeing is dramatically influenced or controlled by the latter process, understanding and predicting this type of diffusion is commercially important. The three approaches most frequently utilized to explain, predict and/or obtain fundamental knowledge about this process are:

(1) Models that adequately explain how various dyes diffuse into different types of fibers.

(2) Methods of measuring the diffusion of dyes into fibers.

(3) Simple and useful calculations of diffusion coefficients of dyes into fibers.

The Fick's law gives a quantitative value to the diffusion of dye molecules from the outside layers into the interior of the fiber. The law receives experimental confirmation when it is applied to an isotropic medium, in particular when there is diffusion through a membrane and the change of concentration on either side of the membrane is small. When considering diffusion into a cylindrical fiber instead of through a membrane, the mathematical calculations become more complex.

3.4.3 Forces Responsible for Fixation

(1) Covalent Bond

The strongest chemical attachment is through the formation of a covalent bond. It applies to reactive dyes mainly.

(2) Ionic Bonds

These bonds play an important part in dyeing protein and polyamide fibers with anionic dyes. In the presence of water or dilute acids the amino group becomes protonated.

(3) Hydrogen Bonds

Both dyestuffs and textile fibers have many groups capable of entering into hydrogen bonding. For example, cellulose contains hydroxyl groups and dyes have azo and amino groups which can make hydrogen bonds with the fiber.

(4) Van der Waals Forces

There are factors in the substantivity of dyes which appear to have no connection with ionic or hydrogen bonds, such as the importance of planar structure, the size of the dye molecule, and the substantivity of dyes towards fiber incapable of forming hydrogen or other bonds, polyesters being an example.

(5) Induced Polarity

It is necessary for the two molecules to be orientated, so that the opposite poles can be adjacent. This need not always be the case, because charges can induce opposite polarity within their sphere of action.

(6) London Dispersion Forces

There is also interatomic attraction between atoms where, apparently, there can be no polarity induced or otherwise, such as in the rare gases. They are called London dispersion forces after their discoverer, and the term dispersion refers to the effect which they have that splitting the spectral band into a number of sub-levels. Atoms or molecules may approach one another in such a way that their respective protons and electrons may temporarily exert an attractive force. This phenomenon can only make one contribution among others which operate to fix the dye on the fiber.

(7) Hydrophobic Interaction

Very briefly, this phenomenon is caused by hydrocarbon groups in an aqueous phase being surrounded by more highly orientated clusters of water molecules (often referred to as "icebergs"). This brings about a decrease in the entropy of the water phase. Removal of the hydrocarbon moiety ofdye molecules results in a gain in entropy because the clusters break down into smaller aggregates of water molecules. The dye which has been absorbed by the fiber will be prevented from migrating back to solution by the hydrophobic interaction.

3.4.4 Sorption of Dyes and Sorption Isotherms

The second process, chemical and/or physical sorption of the dye occurs at equilibrium. This sorption occurs in a manner on the fiber that the dye becomes substantive. Mechanisms by which dyes are absorbed and become substantive to fibers, are dependent on the chemical nature and morphology of the fiber and steric and electronic structure features of dyes. Although there are various methods on explain the sorption behavior of dyes onto fibers, the Nernst, Freundlich and Langmuir sorption isotherms have been proven quite useful in characterizing and classifying dye-fiber sorption processes. These isotherms represent the relationship of the concentration of dye absorbed onto the fiber (C_F) and its concentration in the dyebath (C_S) at a constant temperature. The ratio of the former to the latter is called the partition coefficient K.

Relationship between concentration of dye in fiber (C_∞^F) vs concentration of dye in solution (C_∞^S) described by (a) Nernst, (b) Freundlich and (c) Langmuir sorption isotherms.

In the Nernst isotherm [Figure 3-5(a)], the linear relationship of dye concentration on the fiber and in the bath holds, except for very concentrated dye solutions. When dye concentration on the fiber is plotted against dye concentration in the solution at constant temperature the curve is a straight line which terminates at the point where both the fiber and the dyebath are saturate. There are slight deviations from the linearity of the curve, particularly as the solutions become more concentrate. This system is probably exhibited in its ideal form when dyeing cellulose acetate rayon from an alcoholic dye solution, and it is also essentially true in the case of the application of disperse dyes to polyester fiber in aqueous suspension, because the dyes are all soluble in water only to a very limited extend, and the undissolved particles act as a reservoir to maintain the concentration of the solution.

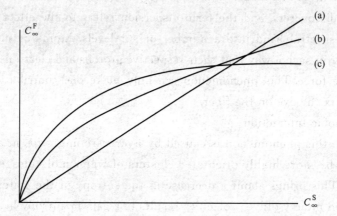

Figure 3-5 Relationship between concentration of dye in fiber (C_∞^F) vs concentration of dye in solution (C_∞^S)

The Freundlich isotherm [Figure 3-5(b)] is representative of the sorption behavior of vat and some direct dyes on the cellulosic fibers. It is characterized by rapid initial dye sorption on the fiber that is limited only by accessibility of fiber surface sites. This limitation occurs because these types of dyes are normally bound to fibers by purely physical forces such as hydrogen bonding and Van der Waals forces. In this case, therefore, adsorption is rapid at first because the sites are easily accessible but becomes slower as the dye molecules have to seek out the more remote points of attachment. The curve, therefore, is not a straight line nor does it reach a point, at which it becomes parallel with the horizontal axis.

The Langmuir sorption isotherm [Figure 3-5(c)] appears to be operative in many ion-exchange processes where the fibers have or acquire a positive or negative charge that dictates attraction of dyes containing an opposite charge. Once the charged fiber sites are occupied on the fiber, sorption of additional dye levels off quite rapidly. The dyeing of wool and nylon with acid dyes, acrylics with basic or cationic dyes, and cellulosic with some direct dyes, they are usually characterized by Langmuir sorption isotherms.

3.4.5 Diffusion Model

The pore model and free volume model are frequently used to explain respectively, the diffusion of dyes into hydrophilic fibers such as cotton and wool, hydrophobic fibers such as polyester. Diffusion of dyes or other, molecules (pore model) occurs only through the liquids in the pores of the polymeric or fiber substrates. These pores are formed by swelling of the amorphous regions of the substrates. The effective diffusion coefficient is thus directly proportional to the porosity, partition coefficient and diffusivity, and inversely proportional to the tortuosity or bending characteristics of the fibers. The glass transition temperature of the polymer that influences diffusion is not considered in this model. To the contrary, the free volume model attributes the diffusion of dyes or other molecules into substrates to the thermal mobility of polymer chains segments. This thermal mobility is directly related to the T_g.

A generalized diffusion model that incorporated feathers of both the pore and free models has been proposed. It purports to explain the diffusion of dyes onto both hydrophilic and hydrophobic fibers. In this generalized model, the most important variables that influence diffusion of dyes into the fiber, enthalpy changes necessary to produce the formation of holes, and a glass transition temperature term that includes constants A and B (derived respectively from the free volume of the polymer chains at T_g and from the coefficient of thermal expansion of the polymer above T_g).

The most frequently employed method of measuring the degree of diffusion of dyes into fibers involves rate of dyeing curves, in which the time of half dyeing is obtained from exhaustion of the dye at equilibrium. When the concentration of the dye on the fiber is plotted against the time or concentration of the dye in the dyebath is plotted against the log of the time, $t_{1/2}$ may be readily computed.

Half-time of dyeing gives valuable information about dyeing properties in practice. If equilibrium is achieved in a comparatively short time, the distribution will be throughout the batch even within the normal process time. However, when equilibrium is reached very slowly under commercial conditions, it is doubtful whether it will be practical to allow sufficient time in the dyebath. Valuable information is also provided about the selection of dyes when more than one is necessary in order to produce the required shade. Obviously, it is desirable, as far as possible, to choose those with about the same half-dyeing time.

New Words and Expressions

convective [kən'vektɪv] adj. 对流的,传送性的	isotherm ['aɪsə(ʊ)θɜːm] n. 等温线
hydrodynamic [ˌhaɪdrəʊdaɪ'næmɪk] adj. 流体动力学的	membrane ['membreɪn] n. 膜,隔膜
aggregation [ˌægrɪ'geɪʃən] n. 聚集(作用),凝聚(作用)	retard ['rɪtɑːd] vt. 阻止,延迟
morphology [mɔː'fɒlədʒɪ] n. 形态学,状态	Zeta potential ζ 电位,电动势
isotropic [ˌaɪsə(ʊ)'trɒpɪk] adj. 等方性的,均匀的	laminar flow 层流
Nernst sorption isotherm 能斯特吸附等温线	Fick's law 菲克定律
Freundlich sorption isotherm 费罗因德利希吸附等温线	pore model 孔道模型

Langmuir sorption isotherm 朗格缪尔吸附等温线	moiety ['mɔɪətɪ] n. [化学]一部分,一半
tortuosity [ˌtɔːtjʊ'ɒsɪtɪ] n. 扭转,弯曲	cluster ['klʌstə(r)] n. 簇,群,簇变形
free volume model 自由体积模型	

Lesson 5　Preparation for Dyeing and Finishing

3.5.1　Desizing

Desizing is the process of removing the size material from the warp yarns in woven fabrics. Most textile materials and fabrics require pretreatments before they can be dyed and finished. If singeing can be considered as the last operation of preparations before chemical processes, desizing is the first of preparatory processes necessary for dyeing and printing.

In the production of woven fabrics, warp yarns are sized with a protective coating to improve the weaving efficiency. Movement of the warp yarn through the heddles and mechanical actions during insertion of filling creates a great deal of abrasive stresses on these yarns. Unprotected, the warp yarns cannot withstand the rigors of weaving. They will break causing machine to stop and thus be responsible for loss of productivity. Weaving efficiencies are vastly improved when the warp is properly sized.

Some of the materials are used either alone or as additives to impart desirable properties to other bases. Starches, polyvinyl alcohol, gums, glue, carboxymethyl cellulose flours, dextrine, polyacrylic acid, gelatins, other synthetic polymers and copolymers can be used as warp sizes. Sizing agents may be divided into two general types: water-soluble, such as gelatin; and water-insoluble, such as starch. Typical synthetic sizes are polyvinyl alcohol, acrylic co-polymers and carboxymethyl cellulose.

When designing the desizing step, it is important to know what base size was used. Each film-former has its own optimum conditions for effective removal. The knowledge of the chemistry of the film-formers will make it easier for one to grasp how to desize specific fabrics.

Historically, starch has been the common choice of film-former for textile sizing. Starch polymers are carbohydrates composed of repeating anhydroglucose units linked together by an alpha glucosidie linkage. The structure contains two secondary hydroxyls at the 2,3 positions and a primary hydroxyl at the 6 position. Highly branched polymers are called amylopectin. Natural starches are not very soluble in cold water. Cooking is necessary to get the starch granules to form a homogenous solution. In addition to natural starch, there are chemical modifications where some of the natural starch properties are altered to make them more useful. Starch used in sizing is often modified to improve its properties and removability in desizing.

Polyvinyl alcohol (PVA) comes in several grades, differing in molecular mass and solution viscosity. Polyvinyl alcohol is manufactured by hydrolyzing polyvinyl acetate. One of the advantages of PVA is that a dried film will redissolve in water without having to degrade it firstly. Fabrics sized with PVA are desized by first saturating with water containing a wetting agent (for rapid penetration) and then by heating in a steamer or J-box (to hydrate the film). Rinsing in hot water to complete the removal completes the desizing step.

Chemical reactions are available for use in desizing. The optimum wash temperature is a function of the grade used to size the warp yarns. Lower molecular mass, partially hydrolyzed grades require lower temperatures than fully hydrolyzed, high molecular mass ones. Temperatures near the boil are required for the fully hydrolyzed grades.

Polyacrylic acid is a water soluble polyelectrolyte that has excellent adhesion to nylon. Therefore, it is used to size filament nylon yarns. The affinity is through hydrogen bonding of the —COOH with amide and amine end groups in the nylon polymer.

Carboxymethyl cellulose (CMC) is made from the reaction of sodium chloroacetate with cellulose. CMC is soluble in cold water and does not require a cooking step. Solutions remain fluid at room temperature and don't retrograde. They can be reheated and cooled repeatedly.

Enzyme desizing, alkali desizing, acid desizing and oxidant desizing are common methods for removing sizes.

3.5.1.1 Enzyme Desizing

Fabrics that contain starch are usually desized with enzymes. Enzymes are specific biocatalysts of organic origin which are produced by living organisms found in living cells and accelerate chemical reactions. Enzymes are named after the compound they break down, for example, amylase breaks down amylose and amylopectin, maltase breaks down maltose and cellulase breaks down cellulose. For desizing starch, amylase and maltase are used. Cellulase, on the other hand, is used for finishing cotton fabrics. Amylase will degrade starch into maltose, a water soluble disaccharide and maltase will convert maltose into glucose, a simple sugar.

Desizing may be done using either batch or continuous methods. Continuous desizing consists of application of the enzyme solution followed by a digestion period to allow the hydrolysis to take place. Finally, the fabric is washed to remove the water soluble digestion products. The digestion period may consist of holding the fabric saturated with enzyme solution at room temperature for several hours. Alternatively, some high temperature stable enzymes can be steamed for a few minutes to heat the fabric and accelerate hydrolysis of the starch.

3.5.1.2 Alkali Desizing

Alkali used as desizing is usually sodium hydroxide, which is found to be used specially for the desizing of cotton or sometimes its blends. Starch takes place strong swelling and becomes loose from adhesion of the fibre under acting of hot sodium hydroxide. Then, by washing effectively, the starch is washed to remove easily. At the same time, the hot alkali can remove part of natural impurities, particularly applicable to cotton fabric containing more cotton shells. Because the alkali used as desizing is mostly waste liquid for scouring, the cost is low and it doesn't damage the fiber. Therefore, the technique is extensively used by dyeing and printing plant. The processing of alkali is shown as follows: padding alkali, piling or steaming, washing.

The recipe for alkali desizing is sodium hydroxide (5-10g/L), wetting agent (1-2g/L), alkali temperature (70-80℃), steaming temperature (>85℃), steaming time (60min), piling time (6-12h).

3.5.1.3 Acid Desizing

The use of acids for the purpose of desizing is not common. Because of thrust of continuous processes and degradation of cotton, acid desizing was displaced. Of course, under proper conditions, dilute sulfuric acid causes a certain degree of decomposition of size and converts it into a product of higher water-solubility; the size is thus easily washed out. When sulfuric and phosphoric acids are applied to fabrics containing starch and cotton, they fail to discern the chemical differences between them. Both are polysaccharides and are subject to many of the same chemical reactions. Mineral acids as well as some organic acids can attack the molecular structures of either starch or cotton.

3.5.1.4 Oxidant Desizing

Oxidative desizing methods using hydrogen peroxide, sodium brominated, ammonium perdisulfate and others can be used to remove starch from cotton fabrics. While there are opportunities to degrade fabrics, the chemical can be controlled so that damages are minimized.

The application of oxidative agents to cause the degradation of starch employs equipment and methods similar to that used for the enzyme process. The pH must be controlled in these processes because oxidation occurs only when the oxygen containing compound is unstabilized to the point that a continuous oxidation potential is achieved.

3.5.2 Scouring

Scouring is a cleaning process used to remove impurities from fibers, yarns or cloth so that they do not interfere with dyeing and finish applications. The amounts and types of impurities and other natural materials present depend on the type of fiber in the material.

Concretely, scouring is an operation in which the quantity of protein, pectin, ash and wax in the fiber is reduced to an amount which will not seriously interfere with subsequent dyeing. It should be regarded as the final purification process and occurs prior to the bleaching step. At the end of scouring, fabrics should be clean and contain the properties of uniform and rapid absorption.

Scouring is accomplished primarily by means of hot alkaline solutions. There are two types of scouring: batch scouring procedure, continuous scouring procedure. The latter is used extensively at present.

3.5.2.1 Batch Scouring Procedure

Kiering is the most important style of batch scouring procedure. It is carried out by means of a kier. Cloth may be scoured in large kettles called kiers. The kier is loaded with cloth in rope form and the scouring chemicals are pumped through the cloth. It is a steel or iron boiler capable of holding 3-5 ton of cloth. The goods are run into the kier in a moist condition and evenly packed. When the kier is filled, the lid is then closed down, the prepared scouring liquor run in and the heating started. The valves are, however, left open until all air has been displaced by steam, for otherwise here is a great danger of damage to the cellulose by the formation of oxycellulose which is somewhat rapidly produced by the action of air on warm cotton impregnated with alkali. The kiering operation depends on the circulation of hot alkaline liquor slightly above the normal boiling point of water that makes it more effective.

3.5.2.2 Continuous Scouring Procedure

The steaming is a key technique of continuous scouring procedure. The continuous scouring procedure can be done using either rope or open width equipment. The rope continuous steaming and open width continuous steaming are its two main methods. They consist mainly of three processes: steeping and pressing of the caustic soda solution, steaming, washing.

(1) Rope Continuous Steaming

When fabrics are to be processed in rope form, the rope commences at the time when fabrics enter the desize J-box and continues through bleaching. After the fabric has been washed following desizing, it is passed through a saturator containing the scour formulation which is maintained at a temperature in excess of 70℃. Elevated temperatures should be maintained so that the process temperature is secured quickly in the J-box. The fabric remains in the J-box for 40-60min, at a temperature of approximately 100%.

Unless care is taken a vast amount of energy would be lost because a large volume of water can transport amounts of energy. The volume of water necessary to remove and suspend the frozen soils must be minimized. When this water volume is at its minimum, the heat requirement is also at its lowest level.

(2) Open Width Continuous Steaming

It should be obvious that when fabrics are scoured in open width, the desize process is also open width. The fabric is processed through a padder or mangle to become saturated with the scouring mixture. Because of the fact that open width steamers provide for short steaming times, chemical concentrations are increased by as much as two times the concentration of that usedfor the rope process.

After the fabric is saturated to at least 100% wet pick-up, the fabric is passed through a steamer. The residence time in these steamers may range from 30 s to 5min. The usual time is 1-2min exposure to wet steam at 100-110℃.

When steaming is completed, fabrics are washed using open washers whose water temperatures are maintained at 70℃ using the counter-flow arrangement.

The effectiveness of the scouring may be assessed by determination of residual wax content, absorbency and mass loss.

3.5.3 Bleaching

Natural fibers, i.e. cotton, wool, linen etc. are off-white in color due to color bodies presented in the fiber. The degree of off-whiteness varies from batch to batch. Bleaching therefore can be defined as the destruction of these color bodies. White is also an important market color so the whitest white has commercial value. Yellow is a component of derived shades. For example, when yellow is mixed with blue, the shade turns green. A consistent white base fabric has real value when dyeing light to medium shades because it is much easier to reproduce shade on a consistent white background rather than on one that varies in amount of yellow.

Bleaching may be the only preparatory process or it may be used in conjunction with other treatments, e.g. desizing, scouring and mercerizing. The combination of such treatments for an individual situation will depend on the rigorousness of the preparation standard and economic factors within the various options. Other chemicals will be used in addition to the bleaching agent. They serve various functions such as to activate the bleaching system to stabilize or control the rate of activation, to give wetting and detergent action, or to sequester metallic impurities. This section gives consideration to the selection of bleaching agents and to the role of the various chemicals used in conjunction.

3.5.3.1 Bleaching with Sodium Hypochlorite

Hypochlorite (ClO^-) bleaching is the oldest industrial method of bleaching cotton. Most cotton fabrics were bleached with sodium hypochlorite up until 1940. It is however the main way of home laundry bleaching products. Sodium hypochlorite is the strongest oxidative bleach used in textile processing. Prior to bleaching with hypochlorite, it is necessary to thoroughly scour fabrics to remove fats, waxes and pectin impurities. These impurities will deplete the available hypochlorite, reducing its effectiveness for whitening fabric. Sodium hypochlorite is the salt of a moderately strong base (NaOH) and a weak acid (HClO). Solutions are therefore alkaline. The species present in a solution can be understood from the following:

$$NaClO \longrightarrow Na^+ + ClO^-$$
$$ClO^- + H_2O \longrightarrow HClO + OH^-$$

(1) Effect of pH

pH has a profound effect on bleaching with hypochlorite.

① If caustic is added to the solution, the equilibrium shifts to the left favoring the formation of the hypochlorite ion (ClO^-) at the expense of hypochlorous acid (HClO).

Under strongly alkaline conditions (pH>10), little or no bleaching takes place.

② When acid is added, the equilibrium shifts to the right and the HClO concentration increases. At a pH between 5 and 8.5, HClO is the major specie present so very rapid bleaching takes place. However, rapid degradation of the fiber also takes place.

③ When the pH drops below 5, chlorine gas is liberated and the solution has no bleaching effectiveness at all.

④ The optimum pH for bleaching is between 9 and 10. Although the concentration of HClO is small, it is sufficient for controlled bleaching. As HClO is used up, the equilibrium conditions continue to replenish it. This pH range is used to minimize damage to the fiber. Sodium carbonate is used to buffer the bleach bath to pH 9 to 10.

(2) Effect of Time and Temperature

Time and temperature of bleaching are interrelated. As the temperature increases, less time is needed. Concentration is also interrelated with time and temperature. Higher concentrations require less time and temperature. In practice, one hour at 40℃ is satisfactory for effective bleaching.

(3) Effect of Metals

Copper and iron catalyze the oxidation of cellulose by sodium hypochlorite degrading the fiber. Fabric must be free of rust spots or traces of metals otherwise the bleach will damage the fabric. Stainless steel equipment is required and care must be taken that the water supply should be free of metal ions and rust from pipes. Prescouring with chelating agents becomes an important step when bleaching with hypochlorites.

Fabrics bleached with hypochlorite will develop a distinctive chlorine odor. This odor can easily be removed with an after treatment consisting of sodium bisulfite and acetic acid. Hypochlorite is used mainly to bleach cellulosic fabrics. It cannot be used on wool, polyamides (nylon), acrylics or polyurethanes (spandex). These fibers will yellow for the formation of chloramides. Bleaching with hypochlorite is performed in batch equipment. It is not used in continuous operations because chlorine is liberated into the atmosphere. Over time, the pad bath decreases in active chlorine causing non-uniform bleaching from beginning to end of the run.

3.5.3.2 Bleaching with Hydrogen Peroxide

Today, it is estimated that 90% to 95% of all cotton and cotton/synthetic fiber blended fabrics are bleached with hydrogen peroxide. It is available commercially as 35%, 50% and 70% solutions. It is a corrosive, oxidizing agent which may cause combustion when allowed to dry out on oxidizable organic matter. Decomposition is accelerated by metal contamination and is accompanied by the liberation of heat and oxygen, which will support combustion and explosions in confined spaces. The material is an irritant to the skin, mucous membranes and dangerous to the eyes.

Hydrogen peroxide is a weak acid and ionizes in water to form a hydrogen ion and a perhydroxyl ion. The perhydroxyl ion is the active bleaching agent.

$$H_2O_2 \longrightarrow H^+ + HOO^-$$

Hydrogen peroxide can also decompose. This reaction is catalyzed by metal ions eg. Cu^{2+}, Fe^{3+}. This reaction is not desired in bleaching because it is an ineffective use of hydrogen peroxide and causes fiber damage.

$$H_2O_2 \longrightarrow H_2O + \frac{1}{2}O_2 \uparrow$$

(1) Effect of pH

Hydrogen peroxide is an extremely weak acid, $K_a = -1.5 \times 10^{12}$. Since the perhydroxyl ion is the desired bleaching specie, adding caustic neutralizes the proton and shifts the reaction to the right. Therefore:

① At pH<10, hydrogen peroxide is the major specie so it is inactive as a bleach.

② At pH 10 to 11, there is a moderate concentration of perhydroxyl ions. pH 10.2 to10.7 is optimum for controlled bleaching. Sodium hydroxide is used to obtain the proper pH.

③ At pH>11, there is a rapid generation of perhydroxyl ions. When the pH reaches 11.8, all of the hydrogen peroxide is converted to perhydroxyl ions and bleaching is out of control.

(2) Effect of Time and Temperature

Stabilized hydrogen peroxide does not decompose at high temperature therefore faster and better bleaching occurs at 95-100℃. This feature makes it ideal for continuous operations using insulated J-boxes or open-width steamers.

(3) Effect of Stabilizers

Stabilizers must be added to the bleach solution to control the decomposition of hydrogen peroxide. Stabilizers function by providing buffering action to control the pH at the optimum level and to complex with trace metals which catalyze the degradation of the fibers. Stabilizers include sodium silicate, organic compounds and phosphates.

Hydrogen peroxide is the bleach most widely used for cellulosic fibers (cotton, flax, linen, jute, etc.) and as well as wool, silk, nylon and acrylics. Unlike hypochlorites, peroxide bleaching does not require a full scour. Residual fats, oils, waxes and pectines do not reduce the bleaching effectiveness of hydrogen peroxide. Additionally, it can be used on continuous equipment. Since it ultimately decomposes to oxygen and water, it doesn't create effluent problems.

3.5.3.3 Bleaching with Sodium Chlorite

Bleaching with sodium chlorite is carried out under acidic conditions which releases chlorine dioxide, a toxic and corrosive yellow-brown gas. Sodium chlorite is sold as an 80% free flowing powder. Chlorine dioxide is thought to be the active bleaching species. It is not used much for bleaching but it is sometimes used to strip dyed goods and is often described as the bleach of last resort. One advantage of sodium chlorite bleaching is that it leaves the fabrics with a soft hand. Because of the gaseous nature, toxicity and corrosiveness of chlorine dioxide, special attention must be paid to the equipment. It must be designed so as to not allow the gas to escape into the work place. Emissions into the atmosphere are of concern, too. The gas corrodes even stainless steel so special passivating treatments must be carried out to prolong the life of the equipment.

When a solution of sodium chlorite is acidified, chlorine dioxide (ClO_2), chlorous acid ($HClO_2$), sodium chlorate ($NaClO_3$) and sodium chloride are formed. Chlorine dioxide and hypochlorous acid are bleaching species, while sodium chlorate and sodium chloride are not. The reactions may be written:

$$ClO_2 + H_2O^- \longrightarrow HClO_2 + OH^-$$
chlorous acid

$$5ClO_2^- + 2H^+ \longrightarrow 4ClO_2 \uparrow + Cl^- + 2OH^-$$
chlorine dioxide

$$3ClO_2^- \longrightarrow 2ClO_3^- + Cl^-$$
chlorate

Chlorine dioxide only reacts with aldehyde groups without affecting hydroxyls or glucosidic linkages. Aldehydes are converted to carboxylic acids. This is of practical importance because cellulose is very slightly damaged, even when high degree of whiteness is obtained. When strong acids are used, the low pH will damage the fiber at the glucosidic linkage so buffers like sodium dihydrogen orthophosphate are commonly used. Sodium acetate and the other sodium phosphate salts are also effective buffers. The acid is added incrementally over the bleach cycle, not all at once. This too controls the bath pH and avoids rapid evolution of chlorine dioxide.

(1) Effect of pH

Chlorine dioxide is favored at low pH 1 to 2.5. It is a more active bleaching agent than hypochlorous acid, which is favored at pH 4 to 5. However, chlorine dioxide is a corrosive and toxic gas. When generated too rapidly, it escapes from the bleaching bath into the atmosphere creating an explosion and health hazard. Once the chlorine dioxide is out of solution, its effectiveness as a bleach is lost.

(2) Effect of Temperature

Little or no bleaching takes place at temperatures below 50℃; however, the bleaching rate increases considerably up to 90℃. Going to the boil is not recommended because it leads to excessive loss of chlorine dioxide with the steam.

3.5.3.4 Optical Brighteners

Certain organic compounds possess the property of fluorescence, which means that they can absorb shorter wavelength light and reemit it at longer wavelength. A substance can absorb invisible ultraviolet rays and reemit them within the visible spectrum. Therefore, a surface containing a fluorescent compound can emit more than the total amount of daylight that falls on it, giving an intensely brilliant white. Compounds that possess these properties are called Optical Brighteners or OBA's. The effect is only operative when the incident light has a significant proportion of ultraviolet rays such as sunlight. When OBA's are exposed to UV fluorescing light bulbs, "black light", the objects glow in the dark, a sure fine way of identifying fibers that are treated with optical brighteners.

3.5.4 Mercerizing

The process of treating cotton fabrics with concentrated solution of sodium hydroxide is called mercerization because it was discovered by John Mercer around 1850. Both Mercerizing and causticizing require cotton to be treated with concentrated solutions of sodium hydroxide (caustic soda). Mercerization requires higher concentrations of caustic soda (19%-26% solutions) whereas causticizing is done with concentrations ranging from 10% to 16%. Both procedures are effective in completing the removal of motes that may have escaped the scouring and bleaching steps. One major difference them is that causticizing improves the dyeing uniformity and dye affinity of cotton without improving luster. Caustic soda solution swells cotton fibers, breaks hydrogen bonds and weakens van der Waals forces between cellulose chains. The expanded and freezed chains rearrange and reorient and when the caustic soda is removed, the chains form new bonds in the reorganized state. When tensionless, the cotton fiber swells, the cross section becomes thicker and the length is shortened. Because of fiber thickening, the fabric becomes denser, stronger and more elastic. Held under tension, the coiled shape of the fiber is straightened and the characteristic lumen almost disappears. The fibers become permanently round and rod like in cross section and the fiber surface is smoother. Decrease in surface area reduces light scattering, adding to fiber luster. Tension increases alignment of cellulose chains which results in more uniform reflection of light. The strength of the fiber is increased about 35%. The fiber also becomes more absorbent. The cellulose crystal unit cell changes from cellulose I to cellulose II and the amorphous area becomes more open, therefore more accessible to water, dyes and chemicals. Mercerized cotton will absorb more dye than unmercerized cotton and in addition, yields an increase in color value at given quantity of dye.

The amount of fiber shrinkage is a measure of the effectiveness of caustic soda's ability to swell cotton. There is the correlation between time, temperature and caustic concentration with fiber shrinkage. It shows that maximum shrinkage occurs with the 24% solution and that most of the shrinkage occurs in the first minute of dwell time. Higher temperatures result in less shrinkage because lower temperatures favor swelling.

3.5.4.1 Chain Mercerizing

Chain mercerizing is done on a range equipped with tenter chains for tension control. The range consists of a pad mangle followed by a set of timing cans and then a clip tenter frame. Fresh water cascades onto the fabric to remove the caustic soda as it is held tensioned in the tenter frame. The length of the frame must match the range speed and assure that the caustic level is reduced below 3% before tensions are released. The tenter frame is followed by a series of open width wash boxes which further reduces the caustic level.

(1) Procedure

① Apply 22% to 25% (48-54℃ Twaddell degrees) caustic at the pad mangle at 100% wet pickup.

② Pass fabric over timing cans. The number of cans must correspond to the range speed and provide at least one-minute dwell time.

③ Clip fabric onto tenter chains and stretch filling-wise while maintaining warp tension.

④ Run fabric under cascade washers to remove caustic. Keep under tension until caustic level is less than 3% otherwise fabric will shrink in filling direction. This width loss is impossible to recover later.

⑤ Release tension and continue washing in open-width wash boxes, to further reduce the caustic.

⑥ Neutralize with acetic acid in the next to last wash box and rinse with fresh water in the last. It is important to control these steps because it is important, in down-stream processing, that the alkalinity remains consistent throughout all production.

(2) Points of Concern and Control

For best results, goods should be dry when entering the liquid caustic impregnation unit. It is necessary to get uniform and even caustic pick-up throughout the fabric. Wet pick-up must be at least 100%. A certain amount of liquid caustic must surround each fiber to provide proper lubrication so that the fibers can be deformed. For piece goods, a caustic concentration between 22%-25% (48-54℃ Tw) should be maintained. Caustic stronger than 25% (54℃ Tw) does not add to mercerized properties whereas below 22% (48℃ Tw), the mercerized fabric will have poor luster and appearance. Caustic solution and impregnated fabric temperatures should be controlled between 21-38℃. Above 38℃, there is a noticeable decrease in luster of the mercerized goods. Below 21℃, there is no noticeable improvement. Proper framing during the washing step is crucial. The goods must be maintained at greige width to one inch over greige for maximum luster. The tensioned width must be maintained throughout the caustic removal operations, otherwise the fabric shrinks and luster is lost. If the optimum washing is obtained there will be only a slight loss in width as the goods come off the tenter clips.

3.5.4.2 Barium Number Test for Mercerization

AATCC Test Method 89 is a common test used for quantifying the degree of mercerization. It is based on the fabric's ability to absorb barium hydroxide. A two-gram swatch of fabric is placed in a flask containing 30mL of a standardized 0.125mol/L barium hydroxide solution. The fabric is stirred for two hours (to allow the barium hydroxide to be absorbed by the fabric). A 10mL aliquot is withdrawn and titrated with 0.1mol/L hydrochloric acid to a phenolphthalein end point. The difference between the starting concentration and the remaining concentration of barium hydroxide is the amount absorbed by the fabric. The procedure is carried out on the fabric both before and after mercerizing and the barium number is calculated as shown below:

Unmercerized fabric will give a barium number of 100 to 105. Completely mercerized fabric will give a barium number of 150. Commercially treated fabrics fall in a range from 115 to 130.

3.5.4.3 Mercerizing Fiber Blends

Color yield, ease of dyeing and uniformity of dyed fabric will offset cost of mercerizing. This holds true even for yarn blends with low levels of cotton. The temptation to mercerize must temper with thoughts about how caustic affects the blending fiber. The following section discusses these issues.

(1) Polyester/Cotton

These can be handled under the same conditions as 100% cotton. Even though polyester fibers are sensitive to caustic, the temperature and time the fibers are in are insufficient to cause fiber damage. One problem with polyester/cotton blends is that they may not be as absorbent as 100% cotton fabrics coming to the caustic saturator. This is because they have, not been given the same thorough scouring and bleaching as 100% cotton. In this case, special penetrating agents are needed to help the caustic solution wet out the fabric.

(2) Cotton/Rayon

Rayon blends pose a number of special problems. Ordinary and high wet modulus viscose rayons are sensitive to caustic solutions. The degree of sensitivity is a function of fiber type and caustic concentration. For example, high wet modulus rayon can withstand caustic better than conventional rayon. Conventional rayon can be dissolved by caustic solutions. High strength caustic solution is less damaging to the high wet modulus rayon than lower strength solutions. Causticizing strength solution will cause the rayon to swell, become stiff and brittle and lose tensile strength. These conditions should be avoided. Fortunately, the higher strengths caustic solution is less damaging so conditions for mercerizing 100% cotton can be used. Special penetrants are also helpful in speeding up the wetting-out process to keep the time rayon is exposed to caustic to a minimum. If conditions are not correct, the damage maybe so severe that the rayon is dissolved.

3.5.4.4 Chainless Mercerizing

Chainless mercerizing is practiced on a range where the cloth is maintained in contact with rotating drums virtually throughout the entire process. The tension on the fabric depends on the friction between the cloth and the surface of the drum. This results in good control of length but limited control of width. Bowed rollers are sometimes used to stretch the width but they are much less effective when compared with the clips of the chain mercerizer. Chainless mercerizing is used on fabrics that cannot be handled on a clip frame such as knits.

3.5.5 Heat Setting

The purpose of heat setting is to dimensionally stabilize fabrics containing thermoplastic fibers. Polyester and nylon are the principal fibers involved. Polyester cotton blended fabrics are produced in large quantities. These fabrics may shrink, or otherwise become distorted either during wet processing or in the consumer's hands. Heat setting is a way of reducing or eliminating these undesirable properties.

The processof passing the fabric through a heating zone for a time and at a temperature that resets the thermoplastic fiber's morphology memory is relatively simple. The new memory relieves the stresses and strains imparted to the fiber by the yarn making and weaving processes, and make the configuration stable it finds itself in flat smooth fabric. This newly imparted memory allows the fiber to resist fabric distorting forces and provides a way to recover from them. The time and temperature needed for the heat treatment depend on fabric density and previous heat treatments. Usually 0-15 seconds at the temperature of 196-213℃ will suffice.

Heat setting reduces polyester's dye uptake. Heat-set goods dye lighter and slower than unheat-set goods. For uniform shades, side to center, front to back and beginning to end exposure to heat must be controlled and uniform, otherwise these differences will show up in the dyed cloth. Heat setting can be done either at the end of wet processing or at the beginning. At either point, the goods must be free of wrinkles and other distortions otherwise the distortions will be permanently set.

There are two main techniques for setting fabrics that are made with man-made fibers: dry heat setting and steam setting. Polyester is usually heat-set dry while nylon may be heat set ether dry or with steam. Continuous heat setting of flat fabrics is usually done with dry heat by contacting the fabric with heated rolls, impinging hot air on the fabric in a tenter frame, a combination of these two methods, or by heating with infrared radiation. Steam, heating setting is often in an autoclave or may be done using continuous steaming equipment. Nylon carpet yarns are often steam set.

Setting is a relative term and if a set fabric is treated under temperature conditions more severe than those at which it was set, the effects of the earlier treatments will be largely remove. It is, however, possible to preset man-made materials and enhance the degree of setting during the dyeing process, as the effects of heat setting treatments are normally additive.

Heat setting can be used to impart a variety of other properties to man-made fibers used alone or in blends, including interesting and durable surface effects such as pleating, creasing, puckering and embossing. Heat setting gives cloth resistance to wrinkling during wear and the ease-of-care properties that may be attributed to improving resiliency and elasticity.

New Words and Expressions

heddle ['hedl] *n.* 综片	piling ['paɪlɪŋ] *n.* 堆置
carboxymethyl [ˌkɑːbɒksɪˈmeθɪl] *n.* 羧甲基	kier [kɪə] *n.* 煮布锅
dextrine ['dekstrɪn] *n.* 糊精	glyceride ['glɪsəˌraɪd] *n.* 甘油酯
anhydroglucose [ænˌhaɪdrəʊˈgluːkəʊs] *n.* 脱水葡萄糖	padder ['pædə] *n.* 浸轧机, 轧车
glucoside ['gluːkˌsaɪd] *n.* 糖苷	mangle ['mæŋgl] *n.* 轧车, 轧液机
amylose ['æməˌləʊs] *n.* 直链淀粉	polyvinyl alcohol 聚乙烯醇
amylopectin [ˌæmmɪləʊˈpektɪn] *n.* 支链淀粉	carboxymethyl cellulose 羧甲基纤维素
pectin ['pektɪn] *n.* 果胶质	ammonium perdisulfate 过硫酸氢铵
J-box J 型箱	sodium hypochlorite 次氯酸钠
bleaching ['bliːtʃɪŋ] *n.* 漂白	chelating agent 螯合剂
chelating ['kiːleɪt] *n.* 螯合	hydrogen peroxide 过氧化氢

bisulfite [ˌbaɪˈsʌlfaɪt] n. 亚硫酸氢盐	sodium chlorite 亚氯酸钠
chloramide [ˈklɔːrəmaɪd] n. 氯酰胺	optical brightener 荧光增白剂
passivate [ˈpæsəveɪt] vt. 钝化	aliquot [ˈælɪkwɒt] n. 可除尽的数
mercerizing [ˈmɜːsəraɪzɪŋ] n. 丝光化	impinge [ɪmˈpɪndʒ] vi. 撞击,接触
causticizing [ˈkɔːstɪsaɪzɪŋ] n. 苛化	autoclave [ˈɔːtəʊkleɪv] n. 高压釜
affinity [əˈfɪnɪti] n. 亲和力	pleat [pliːt] n. 褶裥
orient [ˈɔːrɪənt] v. 取向,定向	crease [kriːs] n. 折皱
tensionless [ˈtenʃənlɪs] n. 无张力	pucker [ˈpʌkə] n. 皱纹,皱褶
lumen [ˈluːmɪn] n. 腔	emboss [ɪmˈbɒs] v. 轧纹,拷花
alignment [əˈlaɪnmənt] n. 队列,结盟	embossing [ɪmˈbɒsɪŋ] n. 凹凸轧花
dwell [dwel] n. 停留	tenter [ˈtentə] n. 拉幅(机)
mercerize [ˈmɜːsəraɪz] vt. 作丝光处理	mote [məʊt] n. 尘埃,微粒
greige [greɪʒ] n. 坯绸,绸坯	

Lesson 6 Dyeing and Printing

3.6.1 Dyeing Method and Process

Textile materials may be dyed through batch, continuous, or semi-continuous processes. The type of process used depends on several things including type of material (fiber, yarn, fabric, fabric construction, garment), generic type of fiber, size of dye lots, and quality requirements in the dyed fabric.

Machinery for dyeing must be resistant to attack by acids, bases, other auxiliary chemicals and dyes. Type 316 stainless steel is usually used as construction material for all parts of dyeing machines that will come in contact with dye formulations.

3.6.1.1 Batch Dyeing Process

Batch processes are the most common method to dye textile materials. Batch dyeing is sometimes called exhaust dyeing because the dyes are gradually transferred from a relatively large volume dyebath to the material being dyed over a relatively long period of time. The dye is said to exhaust from the dye bath to the substrate. Textile substrates can be dyed in batch processes in almost any stage of their assembly into a textile product including fiber, yarn, fabric, or garment. Generally, flexibility in color selection is better and cost of dyeing is lower.

Some batch dyeing machines operate at temperature only up to 100℃. Enclosing the dye machines so that it can be pressurized provides the capability to dye at temperature higher than 100℃. Cotton, rayon, nylon, wool and some other synthetic fibers dye more easily at temperature higher than 100℃.

The three general types of batch dyeing machines are those in which the fabric is circulated, those in which the dyebath is circulated while the material being dyed is stationary, and those in which both the bath and material are circulated. Fabrics and garments are commonly dyed in which the fabric is circulated. The formulation is, in turn, agitated by movement of the material being dyed. Fiber, yarn, and fabric can all be dyed in machines which hold the material stationary and circulate the dyebath. The common used machines in the batch dyeing process include becks, jigs, package dyeing, beam dyeing, skein dyeing, paddle machines, rotary drums, tumblers and jet dyeing. Jet dyeing is the best example of a machine that circulates both the fabric and the dyebath. Jet dyeing machines are excellent for knitted fabrics but woven fabrics may also be dyed using jet machines.

The term "liquor-to-fiber ratio" or just "liquor ratio" is used to express the relative amounts of dyebath and fiber. The liquor ratio is the mass of dyebath used per unit mass of material being dyed. If 1 kilogram of dyebath is used 0.1 kilogram of material being dyed, the liquor ratio is 10 to 1 (10 : 1). Liquor ratio varies over a wide range depending on the type of dyeing process and equipment used. Typical values range from about 50 to 1 for some processes. Low liquor ratio, smaller amount of bath relative to the amount of fiber being dyed, gives higher dyebath exhaustion. Therefore, utilization of dye is usually better under lower liquor ratio dyeing conditions. Many factors determine the optimum liquor ratio for a given dyeing process. Liquor ratio is influential to the utilization of dyes, level dyeing, energy consumption and volume of wastewater. Generally speaking, high liquor ratio is favorable to level dyeing, but it will reduce the efficiency of dyes and increase the wastewater volume. In order to improve the utilization of dye under the circumstances of level dyeing, promoting agent can be used to enhance the dye utilization.

Procedural variations must be closely controlled in dyeing. Obviously, dye and chemical computations and mass must be correct. Reproduction of dyeing cycle time, dyeing temperature, rate of temperature rise, and agitation of the substrate and dyebath must all be controlled for dyeing to be successful.

During the textile fiber production and processing, the textile will be influenced by various tensions, in order to prevent or reduce the contraction and coloring unevenly in the dyeing process, we should eliminate its inner-stress. For example, cotton fabric should be wetted uniformly before dyeing, and synthetic fabric should be heat set.

3.6.1.2 Continuous Dyeing Processes

Continuous dyeing is most suitable for woven fabrics. Most continuous dye ranges are designed for dyeing of blends of polyester and cotton. Nylon carpets are sometimes dyed in continuous processes, but the design of the range for continuous dyeing carpets is much more difficult than that for flat fabrics. Warps can also be dyed in continuous processes. Examples of this are slasher dyeing and long chain warp dyeing usually using indigo.

A continuous dye range is efficientand economical for dyeing long runs of a particular shade. Tolerances for color variation must be greater for continuous dyeing than batch dyeing because of the speed of the process and the large number of process variables that can affect the dye application. The process as showing in Figure 3-6 is often designed for dyeing both the polyester and cotton in a blend fabric in one pass through the range. The polyester fibers are dyed in the first stage of the range by a pad-dry-thermofix process. The cellulosic fibers are dyed in the latter stages of the range using a pad-steam process.

Fabric which was previously prepared for dyeing enters the dye range from rolls. A scray is used to accumulate fabric entering the range so that the range can continue to run while a new roll of fabric is sewn to the end of the strand being run. Uniformity of application of dye requires that continuous dyeing be done in open width. Typical line speed in continuous dyeing is 50 to 150 meters per minute.

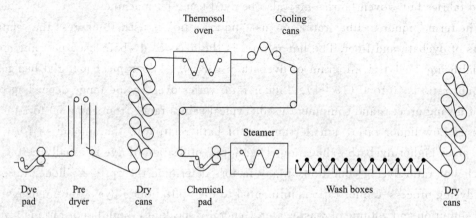

Figure 3-6　Schematic diagram of a continuous dye rang

Padding is a critical step in continuous dyeing. The disperse dye formulation (and sometimes the dye for the cellulosic component) are applied in the first padder. The fabric is immersed in the dye formulation usually at room temperature and squeezed to give a uniform add-on of dye formulation across the width and along the length of the fabric. Low temperature in the formulation in the padder minimizes tailing. Higher temperature promotes wetting of the fabric in the short time the fabric dwells in the pad formulation.

The wet fabric leaving the padder enters a dryer to remove moisture and leave the dye uniformly deposited on the fabric. Radiant predrying using infrared energy inhibits migration of the dye. Drying is completed using steam-heated cylinders.

A thermal treatment called thermosoling fixes the disperse dye on the polyester fiber. The thermosol oven heats the fabric to a temperature of 199-221℃, The exact temperature depending on the particular dyes being applied. The dye sublimes and diffuses into the polyester fibers during the thermosol treatment. The fabric dwells in the thermosol oven for about 1 to 2 minutes.

The cooling cans lower the fabric temperature so that it does not heat the solution in the chemical pad. The chemical padder applies the dyes (and sometimes chemicals) for the cellulosic fibers.

The steamer heats the wet fabric so that the dye can diffuse intothe cellulosic fibers. The fabric usually dwells in the steamer for 30-60 seconds.

The washing section of the range is used for rinses, chemical treatments which may be required to complete the dyeing, and washing of the fabric to remove unfixed dyeand auxiliary chemicals used in the dyeing. The dye and chemical formulation used in the padders and wash boxes depend on the particular classes of dye being applied.

In order to prevent color differences, the wettability of the textile in the continuous dyeing process is supposed to be well. Inconsistent desizing, incomplete desizing material redeposition, and inconsistent scouring are common causes of dye defects. The size may cause a difference in the wettability of the textile material which affects dye uptake. Mercerization has a dramatic effect on dyeability of cellulose and usually improves markedly the dyeability of immature fibers. Since mercerization affects the dyeability of cotton fibers, mercerization must be done uniformly in order for the fabric to be dyed uniformly.

Pad-batch dyeing is a semi-continuous process used mainly for dyeing of cotton fabrics with reactive dyes. Both woven and knitted fabrics can be dyed using this method. Fabric is padded with a formulation containing dye, alkali, and other auxiliary chemicals. The padded fabric is accumulated on a roll or in some other appropriate container and stored for a few hours to give the dye time to react with the fiber. Time, temperature, alkalinity, and reactivity of the reactive dye all influence the process. Pad-batch dyeing is usually done at ambient temperature, but heating of the fabric during batching decreases the time required in the batching stage. Higher alkalinity and selection of more reactive types of reactive dyes also shorten the time required to complete the reaction. Typical batching times range from 4 fours to 24 hours. The batch may be rotated to prevent settling of the formulation and nonuniform dyeing.

Pad-batch techniques have also been applied in preparation of certain types of fabrics for dyeing. Scouring and bleaching of cotton can be done through a cold pad-batch process.

3.6.2 Textile Printing

Textile printing is the most versatile and important methods used for introducing colour and design to textile fabrics. Printing produces localized coloration of textile materials and is often used to produce colored patterns on fabrics or garment. Each color applied in a printing process must be applied in a separate step or position in the printing machine. Printing methods may be classified as direct, discharge and resist.

Direct printing—Dye in a thickened formulation is applied to selected areas of the fabric producing a colored pattern.

Discharge printing—A discharging agent destroys dye on selected areas on a fabric which was previously dyed a solid shade. A white pattern remains where the dye was discharged. Alternatively, a discharge formulation containing dyes that are resistant to discharging produces a second color where the discharge is applied to the previously-dyed fabric.

Resist printing—Dye is applied to a fabric but not fixed. A resist formulation is printed on selected areas of the fabric. The resist agent prevents fixation of the dye in subsequent processing. The unfixed dye is washed away leaving a white pattern. If the resist agent is applied before the dye, the method is called a "preprint process". If the dye is applied first followed by the resist formulation, the method is called an "over-print process".

Printing methods may also be classified according to the process used to produce the pattern. Screen printing and roller printing are the two most common printing methods used in textiles. Ink jet printing, heat transfer printing, and methods based on relief are also practiced in textile.

3.6.2.1 Screen Printing

Silkscreen, or screen printing is a method of applying a colored design created either by hand or by an automated process. In hand screen printing, the fabric to be printed is spread out on a long table. A screen is prepared for each color of the design. Today, screens of synthetic fibers or metal mesh are used for industrial processes.

The printer takes the screen for one color, which is mounted on a frame, and places it in the correct position above and against the fabric. A viscous dye paste, rather than an aqueous dye solution, is used to prevent running of the applied design. The dye paste is coated on the top of the screen. The fabric is underneath. A squeegee is run across the screen and presses the dye through the open area of the screen onto the fabric. After one color has been applied, a second color is added as additional color if they are required.

3.6.2.2 Rotary Screen Printing

A rotary screen printing technique has been developed that makes possible an output of quantities of 25 to 100 meters per minute. Instead of using a flat screen, screens are shaped into cylinders or rollers. The dye feeds through the unscreened area, pushed out from inside the rollers, and up to twelve colors may be printed on one fabric. But the maximum length of the design repeat is limited to the circumference of the roller.

3.6.2.3 Roller Printing

The printing mechanism in roller printing is intaglio. A print roller is engraved with a pattern. Color deposited in the engravings is transferred to the fabric in the printing process (Figure 3-7).

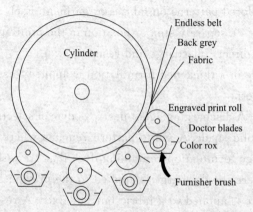

Figure 3-7 Schematic diagram of roller printing process

A furnisher brush picks up print formulation from the color trough and transfers it to the print roll. A doctor blade scrapes color from the smooth, unengraved portions of the print roller and leaves the formulation in the engraved portions. The print formulation in the engraved portions of the print roller is transferred to the fabric. The lint doctor cleans the print roll of lint and trash as well as color picked up by the print roller from previously printed areas of the fabric. The cylinder supports the print blanket which provides a flexible surface which allows the fabric to be compressed into the engraved roll to accept the print formulation. The back grey prevents contamination of the blanket by print formulation which may pass through the fabric.

3.6.2.4 Heat Transfer Printing

Heat transfer printing, or sublimation printing, is a system in which dyes are printed onto a paper base and then transferred the paper to a fabric. The transfer of colors takes place as the color vaporizes or "sublimes". Transfer printing is achieved by rolling or pressing the paper and the fabric together under pressure and at high temperature (200℃).

Equipment for heat transfer printing may he either continuous or batch. Batch processes use a flat bed press to hold the printed paper in contact with the fabric surface while heat is applied. This type of device is common for printing garments and can be used by the garment retailer at the point of sale of the garment.

The machine of continuous heat transfer printing (Figure 3-8) is a calender with feeding devices for the fabric. Hot oil in the cylinder is usually the source of heat. A variation of this machine uses an external infrared heater as the energy source to heat the material and a slight vacuum to hold the fabric in position against the drum. The use of vacuum is said to improve the efficiency of dye transfer from the paper to the fabric.

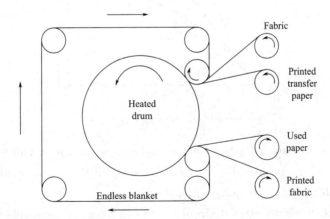

Figure 3-8 Calender type heat transfer printing machine

3.6.2.5 Ink Jet Printing

Ink jet printing has been a staple for office printing for a number of years. It has now made its way into the textile world and is generating much interest in our quick response and just-in-time industry. Current ink jet machines are of two types: continuous ink jet (CIJ) and drop-on-demand (DOD).

3.6.2.6 Printing Formulation

Printing formulations contain color, binders, softeners, thickeners and other auxiliary chemicals. The print paste must be fluid enough to pass through the screen without clogging the holes yet must not flow or wick once the formulation is on the fabric.

Print thickeners are water soluble polymers. Aqueous solutions of these polymers are high in viscosity (resistance to flowing) because of fiction between the large molecules and by tangling and entrapping water in their structure. Under shear, such as stirring or being forced through a small opening, fluids may exhibit any of the behaviors exhibited in Figure 3-9. Shear rate can be thought of as stirring rate or velocity of movement of the fluid. Shear stress can be thought of as the force required to cause the fluid to move or to keep it moving at a certain velocity. A Newtonian fluid is one in which the shear stress is directly proportional to the shear rate.

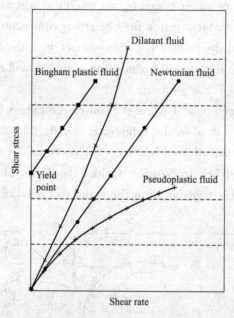

Figure 3-9   Rheolpgy of fluids

Therefore, the viscosity of a Newtonian fluid isconstant with change in shear rate. Non Newtonian fluids, which are pseudoplastic fluids, dilatant fluids and bingham plastic fluids, either increase or decrease in viscosity when the shear rate changes.

Pseudoplastic behavior is often desirable in a print paste. A pseudoplastic print paste will decrease in viscosity and flow easily when it is pushed through the holes in a printing screen. After it is deposited on the fabric, the paste quickly increases in viscosity so that it stays where it was placed on the fabric. Since movement of the color is avoided, the print borders are sharp and well defined.

3.6.2.7 Other Printing Methods

Block printing appears to be the oldest direct printing technique. This is the simplest of the printing techniques and requires only a limited technology. A block of material (usually wood) has a design drawn on one flat side. The design is carved by cutting away the spaces between the areas that form the pattern, thus placing the design in a raised position. Color is applied to the surface of the block, and the block is pressed onto the cloth.

Over-printing is application of a design on a fabric that has already been piece dyed. An overprinted fabric can be distinguished from a blotch print because the overprint will have the same depth of shade on both the face and back. To be successful the color of the printed design should be darker than the background.

Warp printing. Warp yarnscan be printed before they are woven into the fabric. The resulting fabrics have a delicate, shimmery quality that is achieved by the indistinct patterns created in warp printing.

Batik. Batik cloth is made by a wax-resist process. The name batik originates in the Indonesian Archipelago, where resist printing has become an important art form. Wax is applied to the areas that the printer does not want to dye.

Tie dyeing. Resist designs can be produced by the tie dyeing, or tie-and-dye, method in which parts of the fabric are tightly wound with yarns, or the fabric may be tied into knots in selected areas. When the fabric is placed in a dyebath, the covered or knotted areas are protected; tied areas can produce intricate and attractive patterns.

New Words and Expressions

convective [kən'vektɪv] adj. 对流的,传送性的	isotherm ['aɪsə(ʊ)θɜːm] n. 等温线
hydrodynamic [ˌhaɪdrəʊdaɪ'næmɪk] adj. 流体动力学的	membrane ['membreɪn] n. 膜,隔膜
aggregation [ˌæɡrɪ'ɡeɪʃn] n. 聚集(作用),凝聚(作用)	retard ['rɪtɑːd] vt. 阻止,延迟
morphology [mɔː'fɒlədʒɪ] n. 形态,状态	Zeta potential ζ 电位,电动势
isotropic [ˌaɪsə(ʊ)'trɒpɪk] adj. 等方性的,均匀的	laminar flow 层流
Nernst sorption isotherm 能斯特吸附等温线	Fick's law 菲克定律
Freundlich sorption isotherm 费罗因德利希吸附等温线	pore model 孔道模型
Langmuir sorption isotherm 朗格缪尔吸附等温线	free volume model 自由体积模型
batch [bætʃ] n. 间歇式(染色)	jig dyeing 卷染
continuous [kən'tɪnjʊəs] n. 连续式(染色)	package dyeing 筒子纱染色
lot [lɒt] n. 批	jet dyeing 喷射染色
beck [bek] n. 染缸	beam dyeing 经轴染色
jig/jigger [dʒɪɡ]/['dʒɪɡə] n. 卷染机	skein dyeing 绞纱染色
tumbler ['tʌmblə(r)] n. 转笼染色机	paddle machine 桨叶式染色机

scray [skreɪ] n. 小型 J 型堆布箱	rotary drum 转笼式染色机
tailing ['teɪlɪŋ] n. 头尾色差	liquor-to-fiber/ liquor ratio 浴比
predry [priː'draɪ] n. 预烘	slasher dyeing 经轴染色
infrared [ˌɪnfrə'red] n. 红外线	long chain warp dyeing 长链式经纱染色
cylinder ['sɪlɪndə(r)] n. 烘筒,滚筒	pad-dry-thermofix process 轧-烘-熔固工艺
range [reɪndʒ] n. 机组,联合机	pad-steam-process 轧蒸工艺
batching ['bætʃɪŋ] n. 堆置	open width 平幅
flat fabric 平幅织物	add-on 给液量
thermosol dyeing 热溶染色	cooling cans 冷却辊
beck dyeing 绳状染色	ambient temperature 室温
semi-continuous ['semɪkənt'ɪnjʊəs] n. 半连续式（染色）	cold pad-batch process 冷轧堆工艺
design/pattern [dɪ'zaɪn]/['pæt(ə)n] n. 图案	engraving 刻纹
frame [freɪm] n. 网框	back grey 衬布
intaglio [ɪn'tælɪəʊ] n. 凹形雕刻	furnisher brush 给浆辊
engrave [ɪn'greɪv] v. 雕刻	doctor blade 刮浆刀
trough [trɒf] n. 槽	lint doctor 去杂刀
binder ['baɪndə] n. 交联剂	heat transfer printing 热转移印花
thickener ['θɪkənə] n. 增稠剂	dye paste/formulation 色浆
batik [bə'tiːk] n. 蜡染	continuous ink jet (CIJ)连续喷墨
localized coloration 局部着色	drop-on-demand (DOD)按需喷墨
direct printing 直接印花	shear rate 剪切速率
discharge printing 拔染印花	shear stress 剪切应力
resist printing 防染印花	Newtonian fluid 牛顿流体
preprint process 预印工艺	pseudoplastic fluid 假塑性流体
over-print process 罩印工艺	dilatant fluid 胀流性流体
ink jet printing 喷墨印花	Bingham plastic fluid 宾汉塑性流体
screen printing 筛网印花	block printing 模版(凸)印花
rotary screen printing 圆网印花	warp printing 经纱印花
roller printing 滚筒印花	tie dyeing 扎染
sequeegee ['skwiːdʒ] n. 橡皮刮水刷	

Lesson 7 Finishing

A fabric finish, in broad sense, is defined as anything that is done to fabrics after weaving or knitting to change the appearance (what you see), the hand (what you feel), the performance (what the fabric does), and the quality (what it values). All fabric finishing increases the added cost of the fabric.

In narrow sense, finishing may be best regarded as the final stage of the treatment of woven or knitted fabrics to prepare them for the consumer, i.e., the final finishing operations other than such processes as pretreatment, dyeing and printing.

In early days the scope of textile finishing was greatly restricted. The finishes that could be performed were governed by the particular type of fiber and the construction of the fabric. With the appearance of synthetic fibers and new methods of making fibers into fabrics, the finishing technique has reached a new level though much of the finishing is simple in principle. Today there are very few fabrics which are not given some special treatment before they are put on sale.

Grey goods are fabrics, regardless of colours, which have been woven on a loom and have received no wet-or-dry-finishing operations. In many cases the layman would not recognize the grey cloth as the fabric it will become after finishing. Truly, "fabrics are made in the finishing".

For the purpose of clarification, finishes may be divided into three basic categories:
a. Basic finishes;
b. Texturizing finishes;
c. Functional finishes.

The basic finishing processes generally refer to those of stiffening, softening, mangling, etc.

The texturizing finishing processes are used to change the texture of a cloth after it leaves the loom, such as plisse, napping, etc.

The functional finishes are used for extending the functions of a fabric, usually by chemical treatment, such as resisting creasing, water, stains, rot, mildew, moths, bacteria, etc.

Today, major factors influencing the chemical finishing sector may be summarized as follows.
- the globalization of textile manufacture;
- quick response to the demands of the consumer;
- the need for higher quality, higher added value products;

- increased levels of automation and process control in machinery and equipment;
- right-first-time, right-on-time, right-every-time processing;
- greater emphasis on cost reduction by minimizing the use of water, energy and all utilities;
- The challenges posed by environmental issues; e.g. minimizing discharges of chemicals onto land, into air and particularly into water;
- The greater interest in multipurpose chemical finishes.

New Words and Expressions

scope [skəʊp] n. 范围	stiffening ['stɪfnɪŋ] n. 硬挺整理
layman [leɪmən] n. 外行	loom [luːm] n. 织布机
texturizing ['tekstʃəraɪzɪŋ] n. 变形工艺	functional ['fʌŋkʃənl] adj. 功能的
rot [rɒt] n./v. 腐烂	moth [mɒθ] n. 蛀虫
globalization [ˌgləʊbəlaɪ'zeɪʃn] n. 全球化	texturing finish 织物变形整理
issue [ɪʃuː] vt. 流出,发出	napping ['næpɪŋ] n. 拉绒
govern ['gʌvn] v. 决定,支配	mildew ['mɪldjuː] n. 霉,发霉
regardless [rɪ'gɑːdləs] adv. 不管怎样地,不考虑	plisse [plɪ'seɪ] n. 褶裥
functional finish 功能整理	grey cloth 坯布,本色布
integration [ˌɪntɪ'greɪʃn] n. 完整,合而为一	mangling ['mæŋgl] n. 轧光整理

Unit 2 Auxiliary and Additive

Lesson 8 Food Additives

A food additive may be defined as any substance that becomes part of the final food product whether added intentionally or incorporated accidentally. Approximately 2800 substances are used by food processors as additives to food products. Many additional materials find their way into food products in small amounts during the course of growing, harvesting, processing, packing and storage.

3.8.1 Phosphines in Frying Oils and Fats

Poultry, especially chicken, contains a large amount of fatty acids and esters near the surface of the meat, especially in the skin. During frying, the presence of heat and oxygen from the air produces the hydroperoxides of these fatty acids. The hydroperoxides decompose to form polymers, various undefinable gummy materials, aldehydes, acids, ketones, etc., which causes the development of off-flavors and off-odors in the food. The fatty acids and esters which are found in the cooking oil or oxides of fatty acids which likewise decompose and undesirably affect the flavor of foods (chicken) cooked therein.

Monomeric phosphines, and in particular triphenylphosphine, are toxic and not suitable for use in edible fats and oils. Polymeric phosphine compounds are neither digested nor absorbed after oral ingestion. Preferred for use are the polymeric triarylor substituted triarylphosphine compounds with a molecular mass substantially above this range are not effective in preventing hydroperoxide formation. These polymeric phosphine compounds having a molecular mass below about 600 are easily absorbed from the digestive tract of animals and could be metabolized to produce toxic compounds.

Preferred oligomeric and polymeric triarylphosphines and substituted triarylphosphine compounds for use as antioxidants are those derived from polyols. These polyol based triarylphosphine polymers (polyethers) are preferred because they can be made from readily available starting materials. Suitable polyols are glycols, glycerol, sugar alcohols, sugars, including monosaccharides, disaccharides, trisaccharides and tetrasaccharides, among others, and other polyols, for example, pentaerythritol.

Compounds in which the oxygen of the polyether group is replaced with a nitrogen or sulfur are also usable as antioxidants. Hydrocarbyl polymeric phosphine compounds which can be prepared from styryl or alkylene phosphine derivatives are also suitable as antioxidants.

Foods fried in oils containing these polymeric phosphine hydroperoxide inhibiting compounds have an improved flavor over foods fried in oils containing conventional antioxidants, the tocopherols, BHA, BHT, etc. The food in which the fatty acid hydroperoxide formation has been inhibited have a flavor similar to that achieved by frying foods in an oxygen free atmosphere, that is, under a nitrogen or carbon dioxide atmosphere.

These hydroperoxide inhibiting compounds are particularly useful in oils and fats which contain unsaturated fatty acids. Oils which are polyunsaturated, especially safflower oil, sunflower seed oil, soybean oil, and corn oil are particularly susceptible to hydroperoxide formation. Foods fried in these oils, or mixtures thereof, or their corresponding hydrogenated oils, especially need to be protected by these phosphine compounds.

3.8.2 Emulsifiers

Emulsifiers provide low-calorie imitation dairy products that are free of added fat, and thus have substantially reduced caloric contents. Specifically, this process resides in the discovery that the caloric with a partial glycerol ester emulsifier, the major constituent of which is a diglyceride, present in an amount of about 38% to 48%. The triglyceride content is less than the mono-and diglyceride content combined but may be about equal to the diglyceride content, the balance being essentially monoglyceride.

A small amount of an additional hydroxy-containing emulsifier may be employed to obtain hydrophilicity if necessary.

The emulsifier is usable in the amount of preferably 3% to 12% for a whippable topping composition, on a wet basis, which is substantially lower than the fat content normally required for imitation dairy products. At the same time, this process provides compositions having all of the desirable attributes required of conventional imitation dairy products. For instance, in the case of a whippable topping composition, the product has good flavor and eating qualities, good foam stability, good overrun, defined as the ability to incorporate air up to 200% to 300% of the composition initial volume; and good whipping time, defined as the ability to whip to the desired consistency or density, using a household type mixer, in 5 to 10 minutes. Similar results have been obtained with compositions formulated to simulate other dairy products, for instance sour cream, coffee whitener, mellorine, chip dip and cream cheese.

In the case of dry mixes, the compositions are readily reconstituted by admixture with water or milk.

Preferably the plastic partial glycerol ester emulsifier has in addition to about 38% to 48% diglyceride, a diglyceride to monoglyceride ratio of about 5∶1 to about 1.5∶1, the balance being essentially triglyceride.

In a preferred embodiment, the emulsifier is prepared by blending together three partial glycerol ester fractions, a monodiglyceride having a low monoglyceride content, which shall be referred to as the (low mono) monodiglyceride fraction; a soft monodiglyceride; and a smaller amount of a hard monodiglyceride. Up to 10% hard monodiglyceride (based on the total lipid content) can be used although the hard monodiglyceride preferably is employed in amounts as low as about 1% to 2% to obtain the properties desired.

In one embodiment, there was prepared a blend containing three lipid fractions, about 74.2% of a (low mono) monodiglyceride made from 70 I. V. soybean oil having a monodiglyceride content of only about 13%, a diglyceride content of about 43% and a triglyceride content of about 43%; about 24.7% of a soft monodiglyceride having a monoglyceride content of about 40% to 48%, a diglyceride content of 40% to 48% and a triglyceride content of 8% to 12%; and about 1.1% of a hard monodiglyceride having a monoglyceride content of about 40% to 48%, a diglyceride content of 40% to 48% and a triglyceride content of 8% to 12%.

The soft and hard monodiglycerides from SCM Corporation are known as Durem 114 (made from a 75 to 85 I. V. soybean oil, and has a capillary melting point (CMP) of 110 to 125°F) and Durem 117 (made from 5 maximum I. V. soybean oil, and has a CMP of 145 to 150°F). These emulsifier fractions in the proportions stated gave combined mono-, di-and triglyceride contents of about 22%, 43% and 35%, respectively. The lipid blend may contain up to about 1% free glycerin and free fatty acids. In this particular example, the lipid blend had a CMP of about 109°F and an I. V. of about 72.

3.8.3 Dicarboxylic acids and derivatives

Since sorbic acid shows a strong antibacterial force in low toxicity, it is widely used as a preservative agent for drinks and foods. However, sorbic acid is sparingly soluble in water and solubility for water is in the order of 0.16g/100mL at normal temperature, but it dissolves more in water of which the pH value is high by formation of salts. Sorbic acid, however, shows antibacterial action in a form of a free acid and in the case of using as a preservative agent it is preferred to lower the pH value of foods as much as possible. The amount of sorbic acid added varies according to the kind of foods, but in general, is about 0.05% to 0.3% by mass. This addition amount is close to the saturated solubility for water of sorbic acid and it is extremely difficult to perfectly dissolve in foods at low pH values.

As a means for improving the solubility of sorbic acid it is considered to divide it finely. Powdered sorbic acid, however, shows a strong irritating action on the mucous membranes of human beings and it will harm the working environment to make sorbic acid into finely divided powder high in scatterability.

The mucous membrane irritating action of sorbic acid can be avoided by use of its salts, but the antibacterial action inherent in sorbic acid cannot exhibit itself if it is kept in a salt condition.

This process provides sorbic acid-containing powder or granules, free from scatterability and rapidly dissolvable in water, comprising 5% to 90% by mass, preferably 10% to 80% by mass, of finely divided sorbic acid having particle diameter of 50μm or less, 10% to 95% by mass, preferably 20% to 90% by mass, of an easily water-soluble substance which is a solid at normal temperature and 0 to 2% by mass, preferably 0.05% to 1% by mass, of a surface-active agent, characterized by having particle diameter of 300μm or more, preferably 500 to 1500μm.

Such sorbic acid-containing powder or granules, are obtained by drying after making powder or granules having particle diameter of 300μm or more from a mixture consisting of finely divided sorbic acid with particle diameter of 50μm or less, an easily water-soluble substance which is solid at normal temperature, water and/or an aqueous organic solvent and optionally, a hydrophilic surface active agent, or by making powder or granules having particle diameter of 300μm or more from the dry mixture after drying the mixture.

The easily water-soluble substance which is a solid at normal temperature is preferably selected from among additives indispensable for the manufacture of foods. As substances of this kind mention is made of sugars, such as cane sugar, grape sugar, fruit sugar and so on; sugar alcohols, such as sorbitol, mannitol and so on; organic acids, such as citric acid, malic acid, tartaric acid, fumaric acid and so on; salts of organic acids, such as the respective sodium salts or potassium salts of acetic acid, citric acid, malic acid, tartaric acid, fumaric acid and sorbic acid, monosodium glutamate, sodium inosinate and so on, as well as sodium primary phosphate, sodium secondary phosphate, sodium tertiary phosphate, sodium pyrophosphate, acid sodium pyrophosphate, sodium metaphosphate and sodium polyphosphate or their corresponding potassium salts. These substances are preferably selected according to the kind of foods and can be used alone or as a mixture of two members or more.

Sorbic acid is used in finely divided form with particle diameter of 50μm or less so as to be able to immediately dissolve when added to foods.

As the hydrophilic surface active agent, for instance, cane sugar fatty acid esters, preferably those ones which are 11 or more in the HLB, lecithins, preferably high purity lecithins, fatty acid esters of sorbitan (Span-20, for instance), reaction products between sorbitan fatty acid esters and polyoxyethylenes (Tween 20, for instance) and so forth are used alone or as a mixture as the easily, water-soluble substance; in some case, no surface active agents are required.

New Words and Expressions

auxiliary [ɔːgˈzɪlɪərɪ] n. 助剂,添加剂	edible [ˈedəbl] adj. 可食用的
additive [ˈædɪtɪv] n. 添加剂	digestive [daɪˈdʒestɪv] adj. 助消化的
poultry [ˈpəʊltrɪ] n. 家禽肉	diglyceride 甘油二酯
hydroperoxide 氢过氧化物	granule [ˈgrænjuːl] n. 颗粒
decompose [ˌdiːkəmˈpəʊz] v. 分解,腐烂	inosinate 肌苷酸盐
gummy [ˈgʌmɪ] adj. 黏性的	pyrophosphate 焦磷酸盐
odor [ˈəʊdə] n. 气味,臭味	metaphosphate 偏磷酸盐
monomeric [ˌmɒnəˈmerɪk] adj. 单分子构造的	polyphosphate 多聚磷酸盐

triphenylphosphine 三苯基膦	sorbitan 脱水山梨糖醇
metabolize [mə'tæbəlaɪz] vt. 使发生新陈代谢	tocopherol [tɒ'kɒfərɒl] 生育酚,维生素 E
polyols 多元醇	safflower oil 红花籽油
antioxidant [ˌæntɪ'ɒksɪdənt] n. 抗氧化剂	lipid ['lɪpɪd] n. 脂质,类脂
derivative [dɪ'rɪvətɪv] n. 衍生物	sorbic acid 山梨酸
mellorine n. 植物油脂做的冰淇淋	mucous membrane 黏膜
lecithin ['lesɪθɪn] n. 卵磷脂	mannitol 甘露醇

Lesson 9 Auxiliaries in Makeup Products

3.9.1 Perfume

The use of perfume is now extremely popular, and there are hundreds of such products to choose from. The ingredients of a perfume are fragrance molecules, solvents, preservatives, moisturisers, and colours.

Perfumes are made up of fragrance chemicals, most of which have been produced by chemists and these are to be preferred to those produced naturally because they can be absolutely pure and free from possible contaminants of the kind that cause natural fragrances to change over time. Once the molecular structure of a natural fragrance has been deduced then it can be made in the laboratory and maybe even modified slightly to produce a different note.

Composing a perfume is analogous to composing music and the terms are similar. High notes are ones that are most volatile, evaporate quickly, and convey a fresh feeling. Middle notes are less volatile and are often the very heady perfume of oriental flowers. Base notes are the least volatile of all and may have an animalistic smell. Current trends, however, are for lighter note perfumes and these have fewer base notes of the kind that were once taken from animals like the civet cat and the musk deer, although even the molecules associated with those smells can now be synthesized.

To compose a new perfume, you need a blend of many chemicals to achieve a unique fragrance, and the skill of the perfume chemist is to find a combination that evaporates together so that no smell dominates. Moreover, the perfume must continue to smell the same throughout the day and into the evening. This requires the right combination of molecules to be blended. When liquids combine, the rate of evaporation of the resulting mixture will be different from that of their individual components, and yet all must evaporate together to achieve the desired effect. This requires specialist chemical knowledge.

The usual solvent for perfume is alcohol and the final solution consists of 15% of fragrance ingredients dissolved in this solvent. There are more dilute versions, such as eau de toilette, and these are priced accordingly.

3.9.1.1 Fragrances

Perfume here refers to fragrances of which some, in theory, may be secret or at least their relative amounts are. Others have to be identified because some individuals may be sensitive to them and these are alpha-isomethyl ionone, which is described as having a "clean" floral smell. Linalool has a floral smell with a hint of spice. Limonene is found in the peel of oranges. Coumarin has the scent of new-mown hay. Geraniol smells of roses. Citronellol is the main fragrance of geraniums and roses. Isoeugenol smells of vanilla. Citral smells of lemon. Methyl 2-octynoate smells of violets. Eugenol has a spicy smell like cloves. Hydroxycitronellal smells of lime. Farnesol enhances the floral smell of the various perfumes.

3.9.1.2 Preservatives

These are ethylhexyl methoxycinnamate and butyl methoxydibenzoylmethane, which act as UV absorbers to protect the other ingredients.

BHT (butylated hydroxytoluene), benzyl benzoate, benzyl salicylate, and benzyl alcohol are preservatives for the various fragrances. Benzyl salicylate also helps the fragrance molecules to blend together.

Citric acid makes the product slightly acidic and this helps preserve the fragrance molecules, which would be affected by alkaline conditions.

Disodium EDTA is there to sequester any metal ions that might interfere with the other ingredients.

3.9.1.3 Moisturizer

This is butylene glycol dicaprylate/dicaprate, which soothes the skin against any irritation that might be caused by the fragrance molecules.

3.9.2 Toothpaste

Tooth enamel is a form of calcium phosphate known as hydroxyapatite and this protects the dentine, which has nerve fibres. If dentine is exposed because the enamel has been damaged through decay, or the gums have receded, then the teeth become painfully sensitive to cold and hot liquids.

Oral bacteria are present in dental plaque, which is a biofilm that forms on the teeth. The bacteria can convert sugar to acid and thereby corrode the enamel. Plaque can be removed by brushing the teeth with toothpaste that consists of a surfactant and an abrasive. The toothpaste can be made pleasant to taste by the inclusion of an artificial (non-sugar) sweetener.

Converting the hydroxyapatite to fluoroapatite makes it stronger and more acid resistant and this can be done by including fluoride in the toothpaste. In formulations designed for children this needs to be less than in adult formulations because children have a tendency to swallow toothpaste. Adult fluoride toothpaste formulations generally have around 150mg fluoride whereas for children it is half this amount.

Some toothpastes are formulated to plug tiny holes in the enamel.

Dicalcium phosphate dehydrate is a mild abrasive; aqua is water and acts as a solvent; sodium lauryl sulfate is an anionic surfactant; cellulose gum acts to hold all the ingredients together in a smooth paste; aroma is unspecified; sodium monofluorophosphate provides fluoride in the form in which it is present in tooth enamel, and the fluoride component is 100mg; tetrasodium pyrophosphate is an emulsifier and thickening agent; sodium saccharin is an artificial sweetener; sodium fluoride provides some of the fluoride (45mg); calcium glycerophosphate is thought to be more readily absorbed into tooth surfaces than simple phosphate; limonene is a hydrocarbon oil extracted from orange peel.

3.9.3 Hairspray

Hair is mainly made of a protein. As such, it will grow naturally in a way determined by our genes with respect to colour and structure. When we want to change how it looks then we may spend a lot of time and money styling it and, when we have it looking just the way we want it, we need a fixative so that it remains that way. The outer layer of hair can become slightly positively charged and then we find it difficult to control so we need something that will remove static. This is more of a problem for women, more because their hair styles tend to be of longer hair.

The chemical solution to the problem is to spray the hair with a polymer that will coat the strands of hair with a film that hold it in place and act as an antistatic agent, and yet be easily removed when we next wash our hair. We also will want it to be easy to comb and so the hairspray should contain lubricating oil as well.

Alcohol denat is denatured alcohol that contains a bitter tasting chemical to deter its theft and misuse. (By denaturing the alcohol, the company also avoids paying excise duty.) Here it acts as the solvent.

Dimethyl ether is the gas that acts as propellant. It is under pressure in the canister and forces out the content through a nozzle when the pressure is released.

Vinyl butyl benzoate/crotonates copolymer coats the hair to hold it in place and prevent static.

PPG-3 methyl ether is polypropylene glycol polymer modified with methyl groups, and dimethicone copolyol is modified silicone oil with polymer strands attached. These coat the hair, holding it in place and restoring its natural gloss.

Perfume is a mixture fragrances, some natural, some artificial, and it may contain items that some individuals are sensitive to. Hydroxycitronellal smells of lime, limonene smells of oranges, alpha-isomethyl ionone has an open air smell, geraniol smells of roses, citronellol smells of geraniums and roses, coumarin smells of new-mown hay, and amyl cinnamal smells like jasmine.

Everina prunastri/oakmoss extract is extracted from lichen and acts as a fixative for the other fragrances as well as providing a base note to complement them.

Ethylhexyl methoxycinnamate acts to filter out UV rays to protect the spray on the hair.

Aminomethyl propanol is a buffering agent to keep the pH stable.

Benzyl benzoate is a preservative.

There are more than 100 kinds of hairspray, many of which are different versions of popular brands.

New Words and Expressions

antistatic [ˌæntɪˈstætɪk] adj. 抗静电的	animalistic [ˌænɪməˈlɪstɪk] adj. 兽性的
sequester [sɪˈkwestə(r)] vt. 使隔离,螯合	enamel [ɪˈnæml] n. 釉质,珐琅
hairspray [ˈheəspreɪ] n. 发胶,定型剂	volatile [ˈvɒlətaɪl] adj. 挥发性的
makeup [ˈmeɪkʌp] n. 化妆品	evaporate [ɪˈvæpəreɪt] v. 蒸发,挥发
perfume [ˈpɜːfjuːm] n. 香水	butylated hydroxytoluene(BHT) 二叔丁基对甲酚
fragrance [ˈfreɪɡrəns] n. 香精,香料	benzyl benzoate 苯甲酸苄酯
contaminant [kənˈtæmɪnənt] n. 污染物	benzyl salicylate 水杨酸苄酯
analogous [əˈnæləɡəs] adj. 类似的,可比拟的	benzyl alcohol 苄醇
plaque [plæk] n. 牙菌斑	eau de toilette(E.D.T.) 淡香水
sodium lauryl sulfate 十二烷基硫酸钠	ethylhexyl methoxycinnamate 甲氧基肉桂酸乙酸己酯
abrasive [əˈbreɪsɪv] n. 磨料	butyl methoxydibenzoylmethane 丁基甲氧基二苯酰甲烷

Unit 3 Leather Manufacture

Lesson 10 Leather—Extraordinary Product of Nature

As every sports car enthusiast knows, the best models are upholstered in leather, and everyone knows that the very best professional baseball gloves and the sturdiest ski boots for men and women are made of leather. Why do we find leather always in the best? Why do people in the known always insist upon genuine leather? Prestige, durability, eye appeal, and unsurpassed healthful properties are some of the many qualities inherent in this product of nature.

From the earliest civilizations right up to the present time, people throughout the world have held leather in high esteem. Supporting this fact are the large number of leather articles, which archaeologists have unearthed, and the frequent mention of the tanner in the pages of recorded history. Man's high regard for leather is due in no small part to a unique combination of desirable physical properties. But there are reasons beyond this thing which only nature can build into a substance. Nature's products, by their inherent beauty, affect our senses and our emotions. Psychologists call this property the esthetic appeal of a substance. Though it varies in intensity from person to person, an esthetic sense is present in all of us. For example, the desire of a woman to possess a mink coat does not exist simply because it's expensive or is a status symbol; it exists because mink fur appeals to her esthetic nature, her sense of touch and sight, her awareness of beauty. Leather is an extraordinary example of a product that has enjoyed universal appeal throughout the ages, it is indeed the timeless fashion and is destined to continue as such, and just the sound of the leather immediately conjures up pleasant thought and alerts the senses. The scent, the sight, and most of all the feel of a leather article quicken the desire to own it.

In this regard, one of the important terms in a tanner's vocabulary is feeling. He is ever mindful of his obligation to preserve in leather those characteristics which arouse the sense of beauty, and thus to produce leather having a natural feel. It is of course true that imitation products can be printed, prodded, and patterned until some almost look like real leather. But the woman who for years has handled a beautifully aged crocodile leather handbag, or the man who has traveled the world over with a case made of magnificent full grain cowhide, will quickly testify to the fact that products fashioned from leather have no equal.

Time is a very important element in the creation of leather, and further serves to make leather more worthy and attractive to the senses. It first begins with the meticulous care that ranchers give to their herds. It takes nature considerable time—the lifetime of an animal—to slowly and carefully weave together all the meritorious properties that ultimately yield a tanner's raw material. And the advent of new chemicals and modern machinery has shortened the time once required to tan a skin into leather. Once it is produced, we again find that time is on leather's side, for few things possess its enduring qualities. A familiar pair of moccasins is a boy's best friend, and a woman of fashion will wear a lustrous leather coat long after she has replaced most other garments. A beautiful old leather saddle is a source of pride to the horseman who owns one, and even the skier who has perhaps switched to metal skis would not dream of schussing down the trail in anything but a leather boot. And so it goes. It takes nature care, skill and time to create quality goods with appeal. This is why articles made of leather give lasting pleasure and satisfaction.

The process of making leather is unrivaled for pure craftsmanship and chemical ingenuity. In turning back the pages of time, we find that the practice of preserving skins predated recorded history. For this reason, we do not know very much about the methods used by primitive man. As time went on, however, it is quite apparent that tanning grew into a highly developed art. The Egyptians carved pictures into stone over 5000 years ago that depict tanners at work. By the year 500 B.C., the Greeks had developed leather making into a well-established trade. We also know that by this time people had discovered that the bark and leaves of certain trees, if soaked in water, would produce solutions with the capacity to tan. Those who practice it are developed considerable skill in their ability to feel, smell, and taste various noxious solutions to ascertain their strength and suitability.

It is difficult to say exactly when the process of tanning passed from the arts to the sciences. The modern tanner moves in an industry of priceless experience. Steeped in rich traditions, tanning is one of the world's most respected sciences—a process that today has grown into a highly developed technology. The tanner is proud to be associated with a product that transmits its tremendous intrinsic value to the goods that are made from it. As an innovator in process development and as a pace setter in the world of fashion, he stands ready and anxious to produce new leathers to meet the demands of the future.

New Words and Expressions

leather ['leðə(r)] n. 皮革,革制品,皮件	printed leather 印花革
esthetic appeal(sense) 美感	upholstered (椅子等)铺软垫
esteem [ɪ'stiːm] n. 好评	full grain leather 全粒面革
mink coat 貂皮大衣	naked leather 水染革
feeling ['fiːlɪŋ] n. 手感	oil full grain leather 油全粒面革
meticulous [mə'tɪkjələs] adj. 仔细的,精确的	Nubuck 纽巴革
rancher ['rɑːntʃə(r)] n. 牧场(农场)主	boarded leather 搓纹革
meritorious 有价值的,相当好的	embossed leather 压花革

moccasms ['mɒkəsɪnz] n. 软拖鞋,"莫卡辛"鞋	weaved leather 编织革
lustrous ['lʌstrəs] adj. 光亮的,有光泽的	double face leather 双面革
tanner ['tænə(r)] n. 制革工人	chamois leather 油鞣革
buckskin 鹿皮	pull-up leather 变色革
wallet leather 票夹革	crazy horse leather 疯马革
bag and case leather 包袋革	ball leather 球革
fancy leather 美术革	brush off leather 擦色革
pearl leather 珠光革	metallised leather 金属革
patent leather 漆革	aniline leather 苯胺革
corrected grain leather 修面革	sole leather 底革
lining leather 衬里革	antique leather 仿古革

Lesson 11 Leather Technology

Tannersconvert the raw hides and skins into leather. At its simplest, leather is hide or skin which has been treated so that it will not decay, and will last for hundreds of years. Every hide and skin is unique, and varies not only from species to species, but even between individ animals. Using modern techniques, tanners have to produce the relatively uniform leather, and add further features, such as color, softness and fullness, to the leather according to customer's requirements.

Leather industry has a long history. With the development of protein chemistry and synthetic chemistry, it has been firmly established as a technology based on scientific principles. The leather technology has become familiar with a wide range of pure and applied sciences. To produce high quality leather, techoologists must understand the nature of the chemicals used, the way in which they react, the means of controlling this reactivity, and the methods of testing and analyzing the finished product. With this knowledge as a basis, tanners must become familiar with all the practical processes and machinery operation that are necessary to produce leather.

The production in a tannery involves beam-house operations, tanning operation, post-tanning operations and finishing operations.

The following operations are typically carried out in the beam house: soaking, unhairing, liming, fleshing and splitting. Typically, the following operations are carried out in the tanning area: deliming, bating, pickling and tanning. In the tanning process the collagen fiber of the hide and skin is stabilized by the tanning agents so that the hide and skin is no longer susceptible to putrefaction and the tanned hide and skin is called wet-blue. Once the hides and skins are converted to wet-blues, they can be traded as intermediate products. However, if wet-blues are to be used to manufacture consumer products, they need further processing and finishing.

Post-tanning operations generally involve washing out the acids that are still present in the wet-blues. Accorrding to the desired leather the wet-blues are retanned to improve the handle, dyed with water-soluble dyestuffs to produce even color over the leather surface, fatliquored to lubricate the fiber and finally dried. After drying, the leather may be referred to as crust, which is a tradable intermediate product.

Finishing operations are to give the leather as thin a finish as possible without harming the natural characteristic of leather, such as its looks and its ability to breathe. By grounding, coating, seasoning, embossing and ironing, the leather will have a shiny or matt, single or multi-colored, smooth or clearly grained surface. The overall objective of finishing is to improve the appearance of the leather and to provide the appropriate characteristics in terms of color, gloss and handle.

Operations carried out in the beam house, the tanning area and the post-tanning areas are often referred to as wet processing, as performed in processing vessels filled with water to which the necessary chemicals are added to produce the desired reaction. After post-tanning the leather is dried and subsequent operations are referred to as dry processing.

New Words and Expressions

tanner ['tænə(r)] n. 制革工人	tanning agent 鞣剂
hide [haɪd] n. 大皮（特指牛皮、马皮等大型动物的皮）	putrefaction 腐败，腐烂
skin [skɪn] n. 小皮（特指羊皮、猪皮等小型动物的皮）	wet-blue 蓝湿革
decay [dɪ'keɪ] v. 腐朽，腐烂	retan 复鞣
softness ['sɔftnəs] n. 柔软性	handle ['hændl] 手感
fullness ['fulnəs] n. 丰满性	fatliquor 给……乳液加脂
protein ['prəutiːn] n. 蛋白质	crust [krʌst] 坯革，半硝革
pure and applied science 理论科学和应用科学	ability to breathe 透气性能
tannery ['tænərɪ] 制革厂，鞣制车间	grounding ['graundɪŋ] 打底，涂底色
beam-house 准备车间	coating ['kəutɪŋ] 涂饰，涂层
tanning ['tænɪŋ] n. 鞣制	seasoning ['siːzənɪŋ] 喷光亮剂，上光
post-tanning operation 鞣后操作	embossing [ɪm'bɔsɪŋ] 压花，轧花
soaking ['səukɪŋ] 浸水	matt [mæt] 无光泽的，消光的

unhairing [ʌn'hɛrɪŋ] 脱毛	gloss [glɔs] 光亮,光泽
liming ['laɪmɪŋ] 浸灰	processing vessel 加工容器
fleshing ['flɛʃɪŋ] 去肉	cattle ['kætl] 黄牛
splitting ['splɪtɪŋ] 剖层,片皮	buffalo ['bʌfələʊ] 水牛
deliming [dɪ'laɪmɪŋ] 脱灰	saddlery ['sædləri] 马具,马具业
bating ['beɪtɪŋ] 软化	collagen ['kɒlədʒən] n. 胶原蛋白
pickling ['pɪk(ə)lɪŋ] 浸酸	

Unit 4 Pulp and Paper Engineering

Lesson 12 Pulp Process and Pulp End Uses

Pulp consists of wood or other lignocellulosic materials that have been broken down physically and/or chemically such that discrete fibers are liberated and can be dispersed in water and reformed onto a web.

Pulping refers to any processes by which wood (or other fibrous raw material) is reduced to fibrous mass. Basically, it is the means by which the bonds are systematically ruptured within the wood structute. The task can be accomplished mechanically, thermally, chemically, or by combination of these treatments. Existing commercial processes are broadly classified as: chemical, semichemical, semi-mechanical and mechanical pulping. There are in order of increasing mechanical energy required to separate fibers and decreasing reliance on chemical action. As a result, chemical methods rely on the effect of chemicals to separate fibers, whereas mechanical pulping methods rely completely on physical action. The more that chemicals are involved, the lower the yield and lignin content since chemical degrades and solublizes components of the wood, especially lignin and hemicelluloses. On the other hand, chemical pulping yields individual fibers that is not cut and gives strong papers since the lignin, which interferes with hydrogen bonding of fibers, is largely removed.

The typical pulping processes are classified as follows:

Mechanical puling: stone groundwood (SGW) for logs; RMP and TMP for chips.

Chemi-mechanical pulping: chemigroundwood, cold soda, CTMP.

Semichemical pulping: NSSC, high yield sulfite, high yield kraft.

Chemical pulping: kraft, soda and soda-AQ, sulfite (acid and bisulfite).

All chemical pulps must be mechanically worked to develop optimum papermaking properties for various applications. Softwood kraft pulps produced the strongest papers and are preferentially utilized where strength is required. Typical applications are for wrapping, sack, and box-liner papers. Bleached kraft fibers are added to newsprint and magazine grades to provide the sheet with, sufficient strength to run on high-speed printing presses. Bleached grades are also used for toweling and food boards.

Sulfite pulps find a major market in bond, writing, and reproducing papers where good formation and moderate strength are required. Kraft or soda hardwood is usually added for improved formation and opacity. Sanitary and tissue papers also use large amounts of sulfite pulp to obtain the requisite softness, bulk and absorbency.

Mechanical pulp has traditionally been used primarily for newsprint and coated printing grades where it provides a well-filled and formed sheet. Because of improved quality and versatility, markets have opened to a variety of pulps for a wide range of printing grades, tissues, toweling, fluff, coating raw stocks, and food-grade boxboards.

New Words and Expressions

yield [ji:ld] v./n. 得率	soda-AQ 蒽醌碱法
stone groundwood (SGW) 磨石磨木浆	cold soda 冷碱法
refiner mechanical pulp (RMP) 盘磨机械浆,木片磨木浆	wrapping paper 包装纸
thermomechanical pulp (TMP) 热磨机械浆,热磨木片磨木浆	sack paper 纸袋纸
chemigroundwood 化学磨木浆	box-liner paper 纸盒衬里纸
chemithennomechanical pulp (CTMP) 化学热磨机械浆	newsprint ['nju:zprɪnt] n. 新闻纸
neutral sulfite semichemical cooking (NSSC) pulp 中性亚硫酸盐半化学浆	toweling ['taʊəlɪŋ] n. 毛巾纸
food board 食品包装纸板	sanitary papers 卫生纸类
tissue ['tɪʃuː] n. 薄页纸,薄型纸	fluff [flʌf] n. 绒毛浆
coating raw stock 涂布原纸	kraft [kræft] n. 牛皮纸
lignocellulosic [ˌlɪgnəʊˌsekjuˈlɒsɪk] adj. 木质纤维素的	

Lesson 13 Papermaking

Paper derives its name from the reedy plant, papyrus. The ancient Egyptians produced the world's first writing material by beating and pressing together thin layers of plant stem. The first authentic papermaking originated in China as early as 100 A.D., utilizing a suspension of bamboo or mulberry fibers. The Chinese subsequently developed papermaking into a highly skilled art. After a period of several centuries, the art of papermaking extended into the Middle East and later reached Europe, where cotton and linen rags became the main raw materials. Paper was first made in England in 1496. By the end of the 15th century, a number of paper mills existed in Spain, Italy, Germany and France. The first paper mill in North America was established near Philadelphia in 1690.

The development of the paper machine is the most important milestone of the industry. Louis Robert, working at the paper mill owned by Ledger Didot, made his first model of the continuous paper machine in 1796 near Paris and received a French patent for his machine in 1799 at the age of 37. In 1803, a patent was issued to Fourdrinier brothers for the improved continuous paper machine designed by Bryan Donkin. At about the same time, John Dickson, a colleague and friend of Donkin, was working his cylinder machine, which was refined by 1809.

In 1840, groundwood pulping method was developed in Germany. The first manufacture of pulp from wood using soda process was patented on July 1, 1854 to an England inventor named Hugh Burgess. In 1867, a Philadelphia chemist, Benjamin Tilgham, was awarded the U. S. patent for the sulfite pulping process; the first commercial sulfite pulp was produced in Sweden in 1874. C. F. Dahl is credited with the development of the kraft (or sulfate) process. The precursor of the kraft process was originally patented in 1854. A later patent in 1865 covered the incineration of the spent soda liquor to recover most of the alkali used in the process.

These inventions and pionerring prototype provided the basis for the modern paper industry. The twentieth century has seen the rapid refinement and modification of the early and rather crude technology, along with the development of such techniques as refiner mechanical pulping, continuous cooking, continuous multistage bleaching, on-machine paper coating, twin-wire forming, and computer process control.

New Words and Expressions

papyrus [pə'paɪrəs] n. 纸莎草	sulfite pulping process 亚硫酸盐制浆法
beating ['biːtɪŋ] v. 打浆,捶打	kraft (or sulfate) pulping 硫酸盐制浆法
pressing ['presɪŋ] 压榨,压合	refiner mechanical pulping 盘磨机械制浆
papermaking 造纸,抄纸	continuous cooking 连续蒸煮
fiber ['faɪbə] n. 纤维	continuous multistage bleaching 多段连续漂白
paper machine (PM) 造纸机	on-machine paper coating 机内纸张涂布
groundwood pulping 磨木法制浆	twin-wire forming 双网成型
soda process 烧碱制浆法	computer process control 计算机过程控制

Part 4
Applied Literature

Lesson 1　Application Letter

Dear Mr. Chen:

　　Your advertisement in the July 15th of China Daily is of the great interest to me. I'm a graduate from Nanjing Tech University and I majored in Light Chemical Engineering. I have obtained A's all the major subjects and worked as an intern in ABC Textile Company. I loved this job very much and I have made progresses together with my colleagues all the time.

　　Enclosed are my score reports on all relative subjects. I hope that my application will get your favorable consideration.

With best regards,

Yours sincerely,
Li Ming

Lesson 2　Recommendation Letter

Dear Mr. Brown:

　　It gives me a great pleasure to recommend Mr. Cheng as a technician to your department of ABC Textile Company.

　　During academic year 2013-2017, he was a student in our Department, College of Food

Science and Light Industry of Nanjing Tech University. I found him very diligent and intelligent. Worthy of mention also is his personality, honest, reliable, responsible and mature. He often participated in extracurricular activities and contributed a great deal to community affairs.

Though Mr. Cheng graduated from this school one year ago, he keeps and contacts with me very often. I strongly recommend this promising young man and your favorable consideration and assistance to him will be very much appreciated.

With best regards,

Yours truly,
Dr. Zheng

Lesson 3　Resume

Objective
To obtain a challenging position as a buyer of textile with an emphasis in ABC Textile Company
Education
2013.09-2017.06 College of Food Science and Light Industry, Nanjing Tech University, B.E. 2017.09-2020.06 School of Chemistry and Molecular Engineering, East China University of Science and Technology, M.E.
Academic Main Courses
Mathematics Advanced Mathematics Probability and Statistics Linear Algebra Electronics and Computer College English Textile Principium Have a good command of both spoken and written English: Past CET-6; TOEFL 623; GRE 2213
Scholarships and Awards
2014.03　First-class Scholarship 2015.11　RohmHaas Award 2016.11　Academic Progress Award
Qualifications
General business knowledge relating to financial, market Have a passion for the Internet, and an abundance of common sense

Lesson 4 Notification

Dear Mr. /Ms:

Thank you for your letter informing us of Mr. Lee's visit during October 1-7. Unfortunately, Mr. Fang, our manager, is now in Shanghai and will not be back until the second half of October. He would, however, be pleased to see Mr. Lee any time after his return. We look forward to hearing from you.

Best regards,

Yours faithfully,
Mr. Zhang

Lesson 5 Letter of Inquiry

Dear Professor Linc:

I am a professor of College of Food Science and Light Industry, Nanjing Tech University. And I have been studying on Colouration of Textile for many years. I got the information that the 3rd International Commerce Conference on Textile which is to be held in Hilton Hotel in Beijing from Match 10 to 12, 2018, I am very interested in it.

I would like to express my appreciation to you if you could give me the dead line for application and the submission of abstracts and papers.

I shall be looking forward to hearing from you soon.

Best regards,

Yours sincerely,
Dr. Zheng

Lesson 6　Invitation Letter

Dear Professor Zhu:

On behalf of the International Council of Chemical Associations (ICCA), I am very glad to invite you to attend the 10th International Conference on Green and Sustainable Chemistry which will held at Kyoto University in Japan on May 1 to 3, 2018.

The conference will cover developments at fronties of green chemistery and sustainable technology. It will focus on the design, development and implementation of chemical product.

You are an internationally acclaimed scientist at green chemistery. Your participation will be among the highlights of the conference.

We sincerely hope that you could accept our invitation. If you can come, we will send you a cope of the initial program.

We are looking forward your reply.

Best regards,

Sincerely yours,
Dr. Wang

Lesson 7　Meeting Reply

Dear Dr. Wang:

I have received your letter date February 4th, 2018, inviting me to attend the 10th International Conference on Green and Sustainable Chemistry which will held at Kyoto University in Japan on May 1 to 3, 2018.

Thanks for inviting me and I'm happy to accept your invitation, and I will send a copy of my review paper by E-mail before deadline to organizer. I hope it could be arranged for oral presentation at the conference.

Best regards,

Sincerely yours,
Professor Zhu

Lesson 8 Thank-you Note after Meeting

Dear Miss Wang:
  Thank you very much for your warm hospitality during the conference, that make my colleagues and I feel very amiable. The conference was very innovative, that provided us many opportunities of contacting and talking with the scientists from other countries of the world. In addition, the social programs also enable us to know the nice humane scenery.
  We are looking forward to meeting you in Holland in the conference next year.
With best regards,

Yours sincerely,
Dr. Cheng

Lesson 9 Inquiry

Dear Sir/madam:
  We are interested in buying large quantities of grey cloth and silk. We would be obliged if you would give us the quotation per kilogram to Hangzhou. It would also be appreciatd if you could forward samples and your price-list to us. We used to purchase these products from other sources. We may now prefer to buy from your company because we understand that you are able to supply large quantities at more attractive prices. As there is a growing demand for this articles, we have to ask you for a special discount.
  We look forward to hearing from you by return E-mail.
With best regards,

Yours faithfully,
Miss Han

Lesson 10　Offer

Dear Miss Han:

　　Thank you for your letter dated May 11th. We have pleasure in submitting the following quotation for your consideration.

　　We confirm that the prices will remain valid for three months. Packing charges and others duties and taxes are included in the price. You will see that our offer compares favorably with the quotations you can get elsewhere. Since you need large quantities, we shall do it for you at an extra charge 20 percent on the quoted priced. If you agree, we will post the samples in a week.

　　We look forward to receiving your order soon. If you need any further information, please do not hesitate to contact us.

　　Best regards,

Yours faithfully,
Miss Zhou

Appendix

Appendix 1 Grammar and Translation Features of English for Science and Technology

1 人称、时态和语态

1.1 人称

科技英语中常用第三人称或无生命第三人称，因为科技论文的着眼点是客观存在的现象、性质以及实验结果或结论，而非作者本身。但第一人称"we"也经常出现，其含义有两种，一种是在叙述某种实验、现象或结论时，告诉读者我们"we"以前曾报道过；另一种情况是为了引导读者按照作者的思路去考虑问题，这时的"we"已不再是作者自己，而是指作者和读者双方。

We have shown previously that there are quite a few compounds having this structure.

我们在以往的报道中已指出，有好几种化学物具有这种结构。

1.2 时态

科技论文中常用时态是现在时、过去时和现在完成时三种，以被动语态为主。叙述一般的科学原理、公式推导、计算过程和结论时常用现在时；在叙述实验操作或试剂配制时多用过去时。虚拟语气在科技英语中也经常出现，主要用于真实条件句，做科学推理，即在某种条件下进行实验某种现象一定会出现。

$ErCl_3 \cdot 6H_2O$ (99.99%) was purchased from Aldrich and was used without further purification.

$ErCl_3 \cdot 6H_2O$ (99.99%) 购自 Aldrich，使用时未经提纯。

If there were no oxygen in air, fuels would not be able to burn.

如果空气中没有氧气，燃料就不可能燃烧。

If we were to mix 1mol of NaOH with 1L of water, we would find that all of the NaOH would dissolve and dissociate.

如果将1mol NaOH 与1L 水混合，其结果是 NaOH 全部溶解且电离。

1.3 被动语态

科技英语中大约有1/3的动词用于被动语态，因为被动语态可以把所讨论的对象放在主语位置上，更加引人注目。凡是需要着重说明谓语动词和它的动作对象之间的关系、或行为发出者没有必要说明以及难以说明时，往往使用被动语态。而多数被动语句中没有行为的发出者，如果需要表示行为的发出者，则用介词 by 引出。

This is shown in Fig. 2.

如图 2 所示。

As is shown in the illustration…

如图所示……

2 非谓语动词

动词根据其在句中所起的语法作用可分为谓语动词和非谓语动词两种。谓语动词就是动词在句中作谓语的各种形式，它必须与主语的人称和数一致。非谓语动词是指动词在句中不做谓语的形式，它不受主语的人称和数的限制，共有三种：动词不定式、分词和动名词。

特点：(1) 有语态（主动和被动）和时态的变化形式；(2) 可带有自己的宾语、状语和补语等，构成非谓语动词的短语形式。

2.1 动词不定式

动词不定式在动词非谓语形式中应用广泛，在句中可起名词、形容词、副词的作用，主要作主语、宾语、表语、定语、状语和补语 6 种句子成分。

(1) 不定式作主语

句子的谓语必须为单数第三人称形式，而且多数情况下可以在句子主语的位置上加 "It" 作为形式主语。

To apply theory to practice is very important. (＝It is very important to apply theory to practice.)

把理论应用于实践是非常重要的。

(2) 不定式作状语

① 在主语前主要表示目的，译成 "为了、要"。

To understand and use physics, we must have the knowledge of basic mathematics.

为了理解并应用物理学，我们必须具备基础数学知识。

② 在句尾表目的（译成 "来、以便于"）或结果（译成 "从而、以至于"）。

Electrons are lighter than the nucleus and therefore easier to move.

电子比原子核轻，因而容易运动。

Most elements are combined with others to form compounds.

大多数元素能与其他元素化合而形成各种化合物。

(3) 不定式作宾语和表语

This computer needs to be repaired.

这台计算机需要修理。

The purpose of this book is to provide an introduction to physics.

本书的目的在于介绍物理学的初步知识。

They are learning to operate computer.

他们在学习操作计算机。

若动词不定式作句子宾语，而句子本身又有宾语、补语时，一定要在句子宾语的位置上加"it"作形式宾语。但"it"本身无任何词义，不要译成"它"。

We find it difficult to learn mathematics well.

我们感到要学好数学很困难。

（4）不定式作定语

一定要位于被说明的名词后面，作后置定语。它一般表示"将来"或"能力"。

The problem to be considered is the wide application of computers.

要考虑的问题是广泛使用计算机的问题。

This is the best way to solve the problem.

这是解决该问题的最好方法。

（5）特殊句型

"be＋不定式"表示"按计划将要发生的动作"。

We are to discuss this point in detail in the next chapter.

我们将在下一章详细讨论这一点。

（6）let "让、设、令"和 make "使得"后的不定式要省去不定式 "to"。

Electricity makes machines run.

电使机器转动。

2.2 分词

当分词用作动词的非谓语形式时，具有形容词和副词的性质。作为非谓语动词的分词主要有以下两种形式（以及物动词 do 为例）：

现在分词（一般式主动形式）doing

过去分词 done

（1）分词作定语

特点：

① 分词作定语时，与被修饰词之间存在逻辑上的"主谓关系"。现在分词表示主动的主谓关系，过去分词表示被动的主谓关系。

② 分词作定语时只位于被修饰词的前后。一般地，单个分词位于被修饰词之前，也可位于其之后，汉译时没有区别；分词短语一定要位于被修饰词之后。

Electricity is the main power used in industry.

电是工业上使用的主要动力。

Wind is moving air.

风是运动着的空气。

③ 少数情况下，分词（特别是过去分词）作定语时，汉译往往不按定语处理。

Its resistance decreases with the increased temperature.

其电阻随温度上升而下降。（不译成"其电阻随增加了的温度而降低"）

④ 有时，现在分词可表示正在进行的动作，过去分词可表示完成了的动作。

A running machine.

一台正在运转的机器。

An assembled radio.

一台装配好了的收音机。

（2）分词作状语

分词或分词短语作状语时，一定要有逗号与句子的其他部分分开，它通常位于主语前或在句尾。根据主语与分词所表示的动作的主、被动关系而应该使用现在分词或过去分词。

① 分词在主语前作状语时，主要表示时间和条件（可分别汉译成"当……时候"和"如果"），其次是表示原因（汉译成"因为、由于"）。

Flowing through a circuit, the current will lose part of its energy.

当电流流过电路时，要损耗掉一部分能量。

Given current and resistance, we can find out voltage.

若已知电流和电阻，我们就能求出电压。

② 分词或分词短语位于句尾，分词前面多数情况下有逗号，主要表示伴随情况或进一步说明（一般可按词序直译）；其次也可表示结果（译成"因此、这样就"）。

Silver is the best conductor, followed by copper.

银是最好的导体，其次是铜。

（3）分词独立结构

分词作状语时，如果其逻辑主语并不是句子的主语，则可以在分词前加上它自己的逻辑主语，这种带有自己的逻辑主语的分词结构叫做"分词独立结构"。注意，不能把这个分词当成是修饰前面的那个名词的。独立分词结构在句子中只能作状语。它在句中的位置、译法及功能均与一般分词作状语时相似。

Atoms being extremely small, we cannot see them.（处于主语前表示原因）

由于原子极小，所以我们看不见它们。

An electron is about as large as a nucleus, its diameter being about 10-12cm.

电子大约与原子核一样大，其直径约为 10~12cm。（处于句尾作附加说明）

2.3 动名词

动名词的形式与现在分词完全一样，只是语法功能不同，科技文中以动名词的主动式最常见。当动名词带有自己的宾语或状语等时，就称为动名词短语，它在句中作一个整体起某一句子成分的作用。

动名词在句中主要起名词的4种主要作用：主语、宾语、表语和介词宾语，其中以作介词宾语最为常见。

① 动名词作介词宾语

There are many ways of producing electricity.

发电的方法有多种。

② 动名词作主语

动名词作主语与动词不定式作主语一样，句子的谓语动词必为单数第三人称形式，因为它表示一个概念或一件事。

Adding an electron to a neutral atom will give it a negative charge.

把一个电子加到一个中性原子上就会使该原子带负电。

Doing experiments is of great help to our study.

做实验对我们的学习是很有帮助的。

③ 三种常见结构的译法

"by+动名词"：译成"通过……"；

"on（upon）+动名词"，译成"在……以后""一旦……（就）"或"在……时"；

"in+动名词"，译成"在……时候（期间）""在……方面"或"在……过程中"。

By changing the resistance, we can change the current.

通过改变电阻，我们就能改变电流。

Upon receiving a signal, the device will send back a signal in reply.

一旦接收到一个信号，该设备就发回一个信号以作应答。

In preparing this book, they overcame many difficulties.

在编写本书过程中，他们克服了许多困难。

3 从句

长句子是科技英语的另一特点，从化学专业角度比较容易理解这一现象。在一个化学反应中要考虑许多反应条件，诸如浓度、温度、介质、催化剂和时间等。而叙述一个结论时，也必须说明许多条件限定其适用范围。化学专业英语中使用的长句子不一定是很复杂的语法句型，多半是加入条件状语和方式状语，以限定和完善所报道的内容，在翻译时有些状语可以译成形容词以修饰名词。当简单的定语和状语不能完善地表述意思时可以使用从句。主要有状语从句、定语从句和同位语从句等，其中由关系代词 that 和 which 引导的定语从句最为常见。在分析和翻译长句子时，首先要找到主句，然后找到主句中的主语和谓语，最后再看从句和其他成分，不可逐字直译。

3.1 状语从句

特点：

（1）状语从句种类繁多，一般可分为 9 大类，包括时间、地点、原因、方式、条件、目的、结果、比较和让步状语从句。

（2）位置灵活，可位于主句之前或之后，有时也位于主句的中间。汉译时状语从句多数应译在主句前。

（3）状语从句由从属连接词引导，这些连接词大多数有词义，在从句中绝大多数不承担句子成分。

When the material is hot enough, visible light is emitted.

当物质受热到一定程度时会发出可见光。

Before one studies a system, it is necessary to define and discuss some important terms.

在我们学习系统之前，必须定义并讨论一些重要的术语。

（4）as 引导的状语从句

as 可以引导 5 种状语从句，可有 6 种汉语词义。

① 译成"当……时候"（一般表示其从句中的动作与主句动作同时进行）

As the current flows, energy from the battery goes to the resistance.

当电流流动时，电池的能量就给电阻消耗了。

② 译成"随着"（这时主、从句一般存在有表示变化的动词 increase、decrease、rise、fall、change、vary 等）

The force of gravitational attraction between two bodies decreases as the distance between them increases.

两物体之间的万有引力，随着它们之间距离的增加而减弱。

③ 译成"因为、由于"

As air has weight, it exerts force on any object immersed in it.

由于空气具有重量，它对处于其中的任何物体均要施加一个力。

④ 译成"如同"（这时 as 前有表示程度的副词 just、exactly、much、somewhat 等）

Electrons move round the nucleus just as the planets move round the sun.

电子绕原子核旋转，就像行星绕太阳运行一样。

⑤ 译成"如……那样"（用在"as…as"同等比较句型中的第二个 as，这是一个省略式的比较状语从句）

This computer is as heavy as that one (is).

这台计算机与那台一样重。

⑥ 译成"虽然、尽管"［此时从句中的某个词，（科技文中常见的是作表语的形容词）一定要放在 as 前］

Small as they are, atoms are made up of still smaller units.

原子虽小，但却是由一些更小的单元组成的。

⑦ 含 as 的各类词组及固定表达式

as 可作介词，可组成固定词组，也可作状语从句和定语从句的引导词。如其后面只有一个名词或代词，则一般是介词，意为"作为"；若其后是一个句子，则看 as 在从句中是否作了某个句子成分。若 as 在从句中起了主语或宾语的作用，则引导定语从句，可译成"正如……那样"；若 as 在从句中不做句子成分，则引导状语从句。

The current flowing in a circuit is inversely proportional to the resistance in the circuit as long as the applied voltage remains constant.（as long as"只要"；as soon as"一旦……就"；组成复合连接词，引导状语从句）

只要外加电压保持不变，则在电路中流动的电流与该电路中的电阻成反比。

This wire is as long as that one (is).（这里的 as long as 不是固定词组，而是普通的同等比较句型，要通过具体的句子来判断，究竟是固定词组作状语从句的引导词还是一个同等比较的句型）

A computer can be used as well to solve this problem.（as well＝also，副词性词组）

我们也可以用计算机来解这道题。

It is possible to find out the current through this resistor as well as the voltage across it.（as well as＝and，是一个固定词组）

我们能求出流过这个电阻的电流及其两端的电压。

This computer works as well as that one does.（这里的 as well as 是同等比较的句型，而不是一个固定词组，要通过具体的句子来判断）

这台计算机的工作性能与那台一样好。

Mathematics is a very useful tool. As a result, it is widely used in various fields of engineering.（As a result＝therefore，介词短语起插入语作用）

数学是一种非常有用的工具，因此它广泛地用在工程的各个领域中。

As a result of the work of the German physicist Kirchhoff, we are able to analyze a circuit consisting of any number of elements.（As a result of"由于……的结果"，as to＝as for"至于"，组成短语介词，相当于单个介词的作用）

由于德国物理学家基尔霍夫的研究成果，我们现在能够分析由任何个元件所构成的电路。

(5) 注意一词多用现象

有些词既可作引导状语从句的从属连接词，也可作介词或别的词类。如：

Once：副词，意为"曾经""一次"。

连接词，意为"一旦"。

After：介词，意为"在……以后"。

连接词，意为"在……以后"。

副词，意为"以后"。

Since：介词，意为"自从"。

连接词，意为"自从""因为""既然"。

副词，意为"从那时以后"。

While：作引导状语从句的连接词，有多个不同词义，如可表示"在……期间""当……时候""虽然"；而 As 可表示"当……时候""虽然""随着""因为""如同"。

That：可引导状语从句、同位语从句、名词从句或定语从句。

When：可引导状语从句、名词从句或定语从句。

3.2 同位语从句

当一个分句作其前面某个抽象名词的同位语时，该分句就称为那个名词的同位语从句。比如 suggestion（建议）、requirement（要求）、assurance（保证）等。

(1) 结构

the/a/this…＋抽象名词（fact/statement/suggestion/requirement/evidence/advantage/idea/doubt…）＋that 引导的同位语从句

(2) 同位语从句一般只由"that"引导，在从句中不做句子成分，本身无词义。

(3) 译法

① 把同位语从句译在被修饰的抽象名词之前（一般要在从句后加上"这一"两字）

The fact that everything around us is matter is known to all.

我们周围的一切东西均是物质这一事实是大家都知道的。

② 把同位语从句译在冒号后

These experiments serve to confirm Gilbert's conclusion that there are two kinds of electric charge.

这些实验用来证实吉尔伯特的结论：电荷有两种。

③ 采用"动宾"译法，把抽象名词译成动词，同位语从句译成宾语从句

The repulsion between these two charged balls is evidence that bodies having like charges repel each other.

这两个带电球之间的斥力表明具有同种电荷的物体相互排斥。

3.3 定语从句

一个分句在另一个分句中充当了某个名词（或代词）的定语作用时，这种分句称为定语从句。

(1) 位置

定语从句一定要位于被定词后，只能作后置定语（这与汉语中的语序不同）。

(2) 种类

① 限制性定语从句

指直接影响到整个句子确切含义的定语从句，如果去掉该定语从句整个句子的意思就不完整。在形式上，这种定语从句与被定词之间没有逗号存在。

② 非限制性定语从句

只起进一步说明作用的定语从句。在形式上，它与被定词之间通过逗号隔开。

非限制性定语从句又可分为两类：（i）修饰主句中某个名词或代词的非限制性定语从句；（ii）修饰整个主句（或修饰其一部分）的非限制性定语从句。

(3) 引导词

包括关系代词（that、which、whose、who 和 as）和关系副词（where、when、why）。

That：用于人或事物，在从句中主要作主语和宾语。它一般不能引导非限制性定语从句。

Which：只用于事物，在从句中主要作主语、宾语和介词宾语。

Whose：用于人或事物，在从句中只能作定语（译成"其"，表示从属）。

Who：只能用于人，在从句中作主语。

As：用于人或事物，在从句中主要作主语和宾语。

That、which、who、as 等在从句中作主语时，从句谓语的单复数形式完全取决于它们所替代的名词或代词的单复数。

关系副词（where、when、why）在从句中主要作状语。Where 用于表示地点、场合等名词后（如 place、case、situation、point 等），在科技文中 where 经常出现在数学表达式或公式之后，可译为"式中""这里""其中"等。When 用于表示时间的名词（如 time）之后。Why 用于表示理由的名词（如 reason）之后。

(4) 关系词的双重作用

① 关系词在定语从句中一定要作某一句子成分（主要是主语、宾语、介词宾语或状语）。

② 关系词一定要代替前面某个名词或代词（或者整个主句）。关系词本身均无固定的词义，关系词代替哪个词，定语从句就定哪个词。

(5) 关系词的译法

① 若定语从句比较短，汉译时就把定语从句直接放在被定词的前面。关系词可在从句末尾译成"的"。

The factory which was built last year manufactures computers.

去年建造的那个厂是制造计算机的。

② 如定语从句比较长，汉译时可把定语从句单独译成一句，可把关系词译成其所代替的那个词的词义；而非限制性定语从句一般应单独译成一句。

Each mass exerts on the rod a force which is equal and opposite to the force that the rod exerts on it.

每个质量对该棒施加了一个力，这个力与棒对它施加的力大小相等、方向相反。

Rockets perform best in space, where there is no atmosphere to impede their motion.

火箭在太空运行最佳，（因为）那里不存在大气阻碍它们的运动。

(6) 修饰事物时，当先行词是不定代词（all、little、anything、nothing、everything…）以及先行词被序数词、最高级或 only、very、no、any 修饰时，只能用"that"，不能用

"which"。

All that you need to do is press this button.
你只需要按一下这个按钮。
This is the best instrument that we have ever used.
这是我们曾用过的最好的仪器。

(7) 关系代词 which 在定语从句中作介词宾语

① "介词＋which" 在从句中作状语

这种情况在科技文中极为常见，这时 "介词＋which" 必定处于从句句首，定语从句从该介词开始。汉译时将从句译在被修饰的名词或代词之前，而 "介词＋which" 在从句末尾译成 "的" 即可。

The devices on which charges are stored are called capacitors.
储存电荷的器件称为电容器。

② "介词＋which" 在从句中作定语

一般用于非限制性定语从句中，如 one of which（其中之一）；both of which（其两者均）；each of which（其每一个）；some of which（其中有一些）；all of which（它们都）。

Curves have many technical applications, many of which are illustrated in the examples in this chapter.
曲线在技术上有许多应用，其中不少示于本章的例题中。

③ 以 "of which" 开头的定语从句

(i) "of" 是从句中的动词或形容词要求的，这时的 "of which" 在从句中作状语。如 consist of、be composed of、be made of、be made up of、be capable of、be aware of、be sure of 等。

A compound usually shows properties different from those of the elements of which it is made up.
化合物通常所示的性质不同于构成它的元素的性质。

This is the maximum amount of amplification of which the transistor is capable.
这是该晶体管所能提供的最大放大量。

(ii) "of which" 在从句中作定语，修饰从句中的宾语和表语，一般只表示 "其中" 之意。

I am particularly grateful to the editors of the series of which this book is a part.
我特别要感谢本书所属的那套丛书的编辑们。

We can find a few conditions of which only two are necessary (＝only two of which are necessary).
我们能够找到几个条件，其中只有两个是必要条件。

(8) 定语从句中的省略

限定性定语从句中的关系代词和关系副词在某些场合可以省去。在科技文中主要有以下几种：

① 关系代词在从句中作及物动词的宾语时可省去

The unit of voltage we shall use is the volt.
我们将采用的电压单位为伏特。

② 关系代词 which 在从句中作单个介词而非短语介词的介词宾语，而且介词位于从句的末尾

Iron is one of the metals we are most familiar with.

铁是我们最熟悉的金属之一。

③ 在表示时间（time）、方式（way）、原因（reason）、距离（distance）、数量（amount）等名词后的关系副词或"介词＋which"可省去

The reason this circuit is used here is quite clear.（省去了 why 或 for which）

在这里采用这种电路的理由是十分清楚的。

（9）修饰整个主句的非限制性定语从句

① "which" 在从句中作主语，此时从句的谓语只能是单数第三人称形式。

n^2 is even, which indicates that n is even.

n^2 是偶数，这表明 n 是偶数。

② "which" 在从句中作介词宾语的定语，如以 in which case 开头的从句

Atoms may have the same number of protons and a different number of neutrons, in which case they are atoms of the same element, but of different mass.

有些原子可能具有的质子数相同而中子数不同，在这种情况下，这些原子就是属于同一元素的原子，但原子量不同。

（10）由 as 引导的定语从句

① as 引导定语从句时作为关系代词，在从句中一定要承担某一句子成分，主要用作主语或宾语，表示"正如……那样的"。若引导状语从句则为从属连接词，本身在从句中不承担任何句子成分，表示"如同……一样"。

② as 引导非限制性定语从句的特点是修饰整个主句或主句中的某一部分，位置灵活，可以放在主句前、后或中间，但前后一定要有逗号与主句分开。汉译时，多数情况下把它放在主句前。

As the title indicates, this chapter deals with the principle of digital computers.

正如标题所示，本章论述数字计算机的原理。

The material is elastic, as is shown in Fig. 2.

如图 2 所示，这种材料是具有弹性的。

Fluid, as the name shows, is a substance that flows readily.

顾名思义，流体是一种容易流动的物质。

4 省略句

（1）并列复合句中的省略

在并列复合句中，后面分句中与前面分句相同的部分可以省略。

Matters consist of molecules and molecules of atoms.

物质是由分子构成的，而分子是由原子构成的。

（2）在某些状语从句中的省略

状语从句连接词［后面省去了"it is（was）"或"they are（were）"］＋形容词/分词/介词短语/名词等。

The pen point is never allowed to remain in the inkwell while not in use.

在不用时，切勿将笔尖留在墨水池里。

These equations, while fundamental, are less frequently encountered.
虽然这些公式很基本，但并不常遇到。

5 强调句

在科技论文中，句子成分的强调手段主要有以下三种，其中前两种最常见。

(1) 强调句型

It is (was) ＋被强调成分（主语/宾语/状语）＋that/which/who…

可译成"正是……""就是……""是……"（当强调疑问词、连接代词或连接副词时，一般译成"到底、究竟"）。

注意：

① "It"在此不是代词，故不可译成"它"，也不是形式主语。若把"It is (was) that (which/who)"这三个词去掉，剩下的句子构成一个完整的句子（有时句子结构需作调整），表达一个完整的含义，则原来的句子为强调句。

It is the potential difference that causes an electric current to flow in the circuit.
正是电位差使得电流在电路中流动。

② 不论被强调的成分是单数还是复数形式的名词或是别的内容，句型中的"be"均要采用单数形式。根据不同时态，主要是"is"和"was"。

It is the hardware which determines to a large degree the capabilities of the system.
是硬件在很大程度上确定了该系统的能力。（"to a large degree"是介词短语作状语，插在谓语和宾语之间）

It was not until 1925 that the existence of the ionosphere was proved.
直到1925年才证实了电离层的存在。

③ 该句型主要强调句子的主语或宾语，无论强调哪种成分都可以使用"that"；"which"一般只用语强调表示事物的主语或宾语；"who"只能用于强调表示人的主语。

It was Faraday who（也可用 that）first discovered electromagnetic induction.
是法拉第首先发现了电磁感应现象。

It was there that the first satellite was launched.
正是在那儿发射了第一颗卫星。（强调作状语的副词 there，只能用 that）

It is these forces which（也可用 that）lead to magnetic phenomena.
正是这些力导致产生了磁现象。

(2) 用助动词 do (does、did) 强调谓语动词

do/does/did＋动词原形

可译成"的确""确实""实际"等。

Electrons do move from the negative to the positive in a wire.
电子在导线中确实是从负极流向正极的。

(3) 用形容词 very 强调名词

The/this/no/物主代词＋very＋名词，可把 very 译成"正是""就是""最"等。

The very gravity prevents us from flying beyond the atmosphere.
正是这种重力，使我们不至于飞离大气层。

6 虚拟语气

语气分为三种，包括陈述语气（分肯定、否定、疑问三种形式）、祈使语气（表示命令、劝告、请求等，科技文中常用于祈使句的动词有 let、consider、suppose、imagine、take、note、see 等）和虚拟语气。

虚拟语气分三类：表示与事实相反或不可能实现或难以实现的事；表示主观愿望、要求、建议、主张、命令等；表示语气委婉或主观的推测。

科技文中虚拟语气常见形式有如下几种。

（1）用在条件句及主句中

属于第一类虚拟语气，表示与事实相反、或不可能实现或难以实现的事。根据假设条件所涉及的时间，可有三种形式。

假设时间	If 条件句中的谓语形式	主句中的谓语形式
涉及现在（及将来）	用一般过去时 be→were 其他动词用过去式	用过去将来时 should/would/could/might＋动词原形
涉及过去	用过去完成时 had＋过去分词	用过去将来完成时 should/would/could/might＋have＋过去分词
涉及将来	(should＋)动词原形 were＋不定式 表示"若要、万一、一旦"	用过去将来时或一般将来时或一般现在时（用于实现的可能性较大的场合，特别用作给读者的告诫时）

If there were no resistance in the wire, the current would be infinite.

若导线中没有电阻，电流就会无限大。（假设时间涉及现在）

If this new method had been adopted, much time would have been saved.

如果早就采用这种新方法，就会节省好多时间。（假设时间涉及过去，主句和从句都是被动语态）

If the pressure be raised（或 were to be raised）further, the container will break.

若进一步加压，容器就会破裂。（假设时间涉及将来，属于一种告诫）

注意：

① If 条件句中若有 be、have、can（may、should 等）这三类特殊动词时，从属连接词 if 可以省去，这时从句发生部分倒装（其倒装方式与构成一般疑问句的方式相同）。

② 条件从句也可由"provided (that)（若）""in case（假若）""unless（除非）"等引导，还可由介词"without（若没有）""under（在……情况下）""but for（要不是）"等引出的条件状语。

Were there no friction on the ground（＝if there were no friction on the ground），it would be impossible to stop the motion of any body.（也可以写成 It would be impossible to stop the motion of any body were there no friction on the ground）.

若地面没有摩擦，就不可能停止任何物体的运动。

Everything on the earth will lose its weight provided there be no gravity.
若没有地球引力，地球上的一切东西都要失重。

（2）表示主观愿望、要求、建议、主张、命令等

常用句型有：

It is necessary/essential/important/possible/impossible/natural/desirable/…＋that＋（should＋）动词原形……

或者：

It is required/desired/suggested/demanded/proposed/recommended/…＋that＋（should＋）动词原形……

或者：

主语＋require/desire/suggest/demand/propose/recommend/…＋that＋（should＋）动词原形……

It is desired that any machine have no energy loss.
我们希望机器都不要有能量损耗。

Experiments require that accurate measurements be made.
实验要求测量应精确。

（3）表示主观推断或语气委婉

其谓语形式为：should/would/could/might＋动词原形

It would not be difficult to prevent road accidents with the aid of electronic apparatus.
借助电子设备来防止车祸想来并不困难。

Appendix 2　English Word Formation Rules of Common Chemical Names

所谓构词即词的构成，即词在结构上的规律。科技英语构词法的特点如下：

（1）外来语多，很多来自希腊语和拉丁语；

（2）构词方法多，除了非科技英语中常用的三种构词法——转化、派生及合成法外，还普遍采用压缩法、混成法、符号法和字母象形法。

1　Conversion（转化）

由一种词转化为另一种词，叫转化法。例如：

原词	新词	原词	新词
water（$n.$，水）	→water（$v.$，浇水）	dry（$adj.$，干的）	→dry（$v.$，烘干）
charge（$n.$，电荷）	→charge（$v.$，充电）	slow（$adj.$，慢的）	→slow（$v.$，减慢）
yield（$n.$，产率）	→yield（$v.$，生成）	black（$adv.$，向后）	→black（$v.$，后退）

2 Derivation（派生）

通过加前、后缀构成一个新词。

2.1 数目表示法

名称		符号		举例		
tera	T	10^{12}	兆兆			
giga	G	10^9	千兆			
mega	M	10^6	兆	megapascal	MPa	兆帕
kilo	k	10^3	千	kilometer	km	千米
hector	h	10^2	百	hectorgram		百克
deca	da	10	十	decameter		十米
deci	d	10^{-1}	分	decimter	dm	分米
centi	c	10^{-2}	厘	centimeter	cm	厘米
milli	m	10^{-3}	毫	milligram	mg	毫克
micro	μ	10^{-6}	微	micrometer	μm	微米
nano	n	10^{-9}	纳	nanometer	nm	纳米
pico	p	10^{-12}	微微	picosecond	ps	皮秒
femto	f	10^{-15}	毫微微	femto-second	fs	飞秒
atto	a	10^{-18}	微微微	atto second	as	阿托秒

前缀	意义	举例
mono-	单(重)	monocrystal（单晶）
di-	双(重)	carbon dioxide（二氧化碳）
tri-	三(重)	trimethyl aluminium（三甲基铝）
tetra-	四(重)	tetramethyl ammonium chloride（四甲基氯化铵）

2.2 常用词缀

2.2.1 变动词为名词

后缀	意义	举例
-er	……者	generate（发生、产生）→generator（发电机）
-tion	过程,结果,状态	operate（操作）→operation（操作或运转结果）
-sion		inject（注射）→injection（注射过程或结果）
-ment		treat（处理）→treatment（治疗、处理）
-ure		press（压）→pressure（压力、强制）
-sis		analyse（分析）→analysis（分析）
-ing	动词＋ing	thicken（增稠）→thickening（增稠过程）

2.2.2 变名词、形容词为动词

后缀	意义	举例
-fy	……化	liquid（液体）→liquidify（液化）
-ize		optimum（最佳）→optimumize（最佳化）
-ate	使……起来	active（活跃的）→activate（使活跃起来）

2.2.3 变形容词为名词

后缀	意义	举例
-ity	性质、状态	viscous（黏的）→viscosity（黏性）
-cy	性质、状态	frequent（频繁的）→frequency（频率）
-ce		present（现存的）→presence（存在）
-ness	性质、状态	thick（厚的）→thickness（厚度）

2.2.4 变名词仍为名词

后缀	意义	举例
-ist	……学家、者	science（科学）→scientist（科学家）
-er	……学家、者	astronomy（天文学）→astronomer（天文学家）
-or		library（图书馆）→librarian（图书管理员）
-ian		mathematics（数学）→mathematician（数学家）
-ician		vegetable（蔬菜）→vegetarian（素食主义者）
-arian		
-age	……量、……数	dose（剂量）→dosage（用量）

2.2.5 变名词为形容词

后缀	意义	举例
-ous	……的	pore（孔）→porous（多孔的）
-ic	……的	graph（图）→graphic（图解的）
-tic		analyse（分析）→analytic（分析的）
-ful	有……的	power（力量）→powerful（有力的）
-able	可……的	solve（溶解）→solvable（可溶解的）
-al	……的	function（功能）→functional（功能的）
-ar	……的	pole（极）→polar（极的）
-ive	……的	act（行动、动作）→active（活动的）
-ile	……的	infant（小儿）→infantile（小儿的）
-y	……的	milk（乳）→milky（乳状的）
-ed	成……的	powder（粉）→powdered（变成粉的）
-lent	多……的	virus（病毒）→virulent（剧毒的）
-scent	产生、发展的过程	lumin［(拉丁语)光］→luminescent（发光的）

3 Composition（合成法）

由两个已有的词合成一个新词。

形式	单词	意义
副词+过去分词	well-known	著名的
名词+名词	carbon steel	碳钢
	rust-resistance	防锈
名词+过去分词	computer-oriented	面向计算机的
介词+名词	by-product	副产物
动词+副词	make-up	化妆品
	clean-up	检查
形容词+名词	atomic weight	原子量
	periodic table	周期表
动词+代词+副词	pick-me-up	兴奋剂
副词+介词+名词	out-of-door	户外

4 Shortening（压缩法）

只取词头字母，如：

单词	来源	中文译文
TOEFL	Test of English as a Foreign Language	非英语国家英语水平考试
ppm	parts per million	百万分之一

将单词删去一些字母，如：

单词	来源	中文译文
lab	laboratory	实验室
kilo	kilogram	千克
flu	influenza	流行性感冒

5 Blending（混成法）

把两个词的一头一尾连在一起，构成一个新词，如：

单词	来源	中文译文
positron	positive（正的）+electron（电子）	正电子
medicare	medical（医学的）+care（照管）	医疗保障
aldehyde	alcohol（醇）+dehydrogenation（脱氢）	醛

Appendix 3 Names and Abbreviations of Common Polymers

缩写	英文名称	中文名称
AAS	Acrylonitrile-acrylate-styrene copolymer	丙烯腈-丙烯酸酯-苯乙烯共聚物
ABS	Acrylonitrile-butadiene-styrene copolymer	丙烯腈-丁二烯-苯乙烯共聚物
ALK	Alkyd resin	醇酸树脂
AMMA	Acrylonitrile-methylmethacrylate copolymer	丙烯腈-甲基丙烯酸甲酯共聚物
AMS	Alpha methyl styrene	α-甲基苯乙烯
AS	Acrylonitrile-styrene copolymer（see SAN）	丙烯腈-苯乙烯共聚物
ASA（AAS）	Acrylonitrile styrene-acrylate copolymer	丙烯腈-苯乙烯-丙烯酸酯共聚物
BMC	Bulk moulding compound	团状模塑料
CA	Cellulose acetate	乙酸纤维素
CAB	Cellulose acetate butyrate	乙酸-丁酸纤维素
CAP	Cellulose acetate propionate	醋酸-丙酸纤维素
CF	Casein formaldehyde resin	甲酚-甲醛树脂
CFE	Polychlorotrifluorethylene（see PCTFE）	聚三氟氯乙烯
CM	Chlorinated polyethylene（see CPE）	氯化聚乙烯
CMC	Carboxymethyl cellulose	羧甲基纤维素
CN	Cellulose nitrate	硝酸纤维素
COPE	Polyether ester elastomer	聚醚酯弹性体
CP	Cellulose propionate（CAP）	丙酸纤维素
CPE	Chlorinated polyethylene（PE-C）	氯化聚乙烯
CPVC	Chlorinated polyvinyl chloride（PVC-C）	氯化聚氯乙烯
CS	Casein plastics	酪素塑料
CTA	Cellulose triacetate	三乙酸纤维素
DMC	Dough moulding compound	圈状模塑料
E/P	Ethylene propylene copolymer	乙烯-丙烯共聚物
CA-MPR	Elastomer alloy melt processable rubber	弹性体合金-可熔融成型橡胶
EA-TPV	Elastomer alloy thermoplastic vulcanizate	弹性体合金-热塑性硫化料
EC	Ethylene cellulose	乙基纤维素
EEA	Ethylene ethylacrylate copolymer	乙烯-丙烯酸乙酯共聚物
EP	Epoxide or epoxy（cured）	环氧树脂
EPDM	Ethylene propylene diene terpolymer	三元乙丙共聚物

续表

缩写	英文名称	中文名称
EPS	Expandable polystyrene	可发性聚苯乙烯
ETFE	Ethylene/tetrafluoroethylene	乙烯-四氟乙烯共聚物
EVA	Ethylene vinyl accetate copolymer	乙烯-乙酸乙烯酯共聚物
EVAL/EVOH	Ethylene vinyl alcohol copolymer	乙烯-乙烯醇共聚物
FEP	Fluorinated ethylene propylene (TFE-HEP)	四氟乙烯-六氟丙烯共聚物
CPMC	Granular polyester moulding compound	粒状聚酯模塑料
HDPE	High density polyethylene (PE-HD)	高密度聚乙烯
HIPS	High impact polystyrene (TPS or IPS)	高抗冲聚苯乙烯
HMWPE	High molecular weight polyethylene	高分子量聚乙烯
LCP	Liquid crystal polymer	液晶聚合物
LDPE	Low density polyethylene (PE-LD)	低密度聚乙烯
LLDPE	Linear low density polyethylene	线型低密度聚乙烯
MMBS	Methyl methacrylate-butadiem-styrene	甲基丙烯酸甲酯-丁二烯-苯乙烯
MC	Methyl cellulose	甲基纤维素
MDPE	Medium density polyethylene (PE-MD)	中密度聚乙烯
MF	Melamine formaldehyde resin	三聚氰胺-甲醛树脂
MPF	Melamine phenol formaldehyde resin	三聚氰胺-酚甲醛树脂
NC	Nitroncellulose	硝基纤维素
PA	Polyamide or nylon	聚酰胺(尼龙)
PA6	Polyamide 6 or nylon 6	尼龙 6
PA11	Polyamide 11 or nylon 11	尼龙 11
PA12	Polyamide 12 or nylon 12	尼龙 12
PA46	Polyamide 46 or nylon 46	尼龙 46
PA66	Polyamide 66 or nylon 66	尼龙 66
PA610	Polyamide 610 or nylon 610	尼龙 610
PAA	Polyacrylic acid	聚丙烯酸
PAA6	Polyaryl amide	聚芳酰胺
PAN	Polyacrylonitrile	聚丙烯腈
PB	Polybutene-1	聚-1-丁烯
PBI	Polybenzimidazole	聚苯并咪唑
PBT	Polybutylene terephthalate	聚对苯二甲酸丁二醇酯
PC	Polycarbonate	聚碳酸酯(防弹胶)
PCTFE	Polychlorotrifluorethylene	聚三氟氯乙烯
PDAP	Polydiallyl phthalate	聚苯二甲酸二烯丙酯
PE	Polyethylene	聚乙烯
PEBA	Polyether block amide	酯-酰嵌段共聚物
PEC	Chlorinated polyethlene (see CPE)	氯化聚乙烯

缩写	英文名称	中文名称
PEEK	Polyether ether ketone	聚二醚酮
PEEL	Polyether ester (YPBO)	醚酯
PEI	Polyether imide	聚醚酰亚胺
PEK	Polyether ketone	醚酮
PEKK	Polyethylene ketone ketone	聚乙烯酮酮
PEOX	Polyethylene (oxide)	聚氧化乙烯
PES	Polyether sulphone	聚醚砜
PET	Polyethylene terephthalate	聚对苯二甲酸乙二醇酯
PETP	(see PET)	
PF	Phenol formaldehyde resin	酚醛树脂
PI	Polyimide	聚酰亚胺
PMC	Polyester moulding compound	聚酯模塑料
PMCA	Polymethy α-chloacrylate	聚α-氯化丙烯酸甲酯
PMI	Polymethacrylimide	聚甲基丙烯酰亚胺
PMMA	Polymethyl methacrylate (acrylic)	聚甲基丙烯酸甲酯
PMMA-T	Toughened acrylic	钢化丙烯酸
PO	Polyolefine	聚烯烃
POM	Polyoxymethylene or Acetal or polyformaldehyde	聚甲醛
POM-CO	Acetal copolymer	聚甲醛共聚物
POM-H	Acetal homopolymer	聚甲醛均聚物
PP	polypropylene	聚丙烯
PPC	Chlorinated polypropylene	氯化聚丙烯
PPE	Polyphenylene ether (see PPO)	聚苯醚
PPO	Polyphenylene oxide-usually modified	聚苯醚
PPOX	Polypropylene oxide	聚氧化丙烯
PPS	Polyphenylene sulphide	聚苯硫醚
PPSU	Polyethylene sulphone	苯砜
PPVC	Plasticized polyvinyl chloride (PVC-P)	增塑聚氯乙烯
PS	Polystyrene (GPPS)	聚苯乙烯
PSU	Polysulphone resin	聚砜树脂
PTFE	Polytetrafluoroethylene	聚四氟乙烯
PU	Hard polyurethane elastomer	硬聚氨酯弹性体
PUR	Polyurethane	聚氨酯
PVAC	Polyvinylacetate	聚乙酸乙烯酯
PVAL	Polyvinylalcohol	聚乙烯醇
PVB	Polyvinylbutyral	聚乙烯醇缩丁醛

续表

缩写	英文名称	中文名称
PVC	Polyvinyl chloride	聚氯乙烯
PVCC	Chlorinated polyvinyl chloride	氯化聚氯乙烯
PVDC	Polyvinylidene chloride	聚偏二氯乙烯
PVDF	Polyvinylidene fluoride	聚偏二氟乙烯
PVF	Polyvinyl fluorde	聚氟乙烯
PVFM	Polyvinylformal	聚乙烯醇缩甲醛
PVK	Polyvinyl carbazole	聚乙烯基咔唑
PVP	Polyvinyl pyrrolidone	聚乙烯吡咯烷酮
SAN	Styrene acrylonitrile copolymer	苯乙烯-丙烯腈共聚物(透明大力胶)
SBS	Styrene butadiene styrene block copolymer	苯乙烯-丁二烯-苯乙烯嵌段共聚物
SEBS	Styrene butadiene styrene block copolymer (saturated)	苯乙烯-丁二烯-苯乙烯嵌段共聚物(饱和)
SI	Silicone	聚硅氧烷
SMA	Styrenemaleic anhydride copolymer	苯乙烯-马来酐共聚物
SMC	Sheet moulding compound	模压塑料板材
SMS	Styrene methylstyrene copolymer	苯乙烯-甲基苯乙烯共聚物
TP-EE	Thermoplastic elastomer-ether ester	醚酯类热塑性弹性体
TP-EPDM	Thermoplastic elastomer-based on EPDM	三元乙丙橡胶类热塑性弹性体
TP-EVA	Thermoplastic elastomer-based on EVA	乙烯乙酸酯共聚物类热塑性弹性体
TP-NBR	Thermoplastic elastomer-based on NBR	丁腈橡胶类热塑性弹性体
TPE	Thermoplastic elastomer rubber	热塑性弹性体
TPO	Thermoplastic polyolefin	热塑性聚烯烃
TPO-XL	Thermoplastic polyolefin rubber-crosslinked (rubber)	交联热塑性聚烯烃
TPR	Thermoplastic rubber (etaslomer)	热塑性橡胶
TPU	Thermoplastic polyurethane	热塑性聚氨酯
TPV	Thermoplastic elastomer or rubber-crosslinked (rubber)	交联热塑性弹性体或橡胶
UF	Urea formaldehyde resin	脲醛树脂
UHMWPE	Ultrahigh molecular weight polyethylene	超高分子量聚乙烯
UP	Unsaturated polyester resin	不饱和聚酯树脂
UPVC	Unplasticized polyvinyl chloride	未增塑聚氯乙烯
VC/E	Vinylchloride ethylene copolymer	氯乙烯-乙烯共聚物
VC/E/MA	Vinylchloride ethylene maleic acid copolymer	氯乙烯-乙烯-马来酸共聚物
VC/E/VAC	Vinylchloride ethylene vinyl acetate copolymer	氯乙烯-乙烯-乙酸乙烯酯共聚物
VC/MA	Vinylchloride maleic acid copolymer	氯乙烯-马来酸共聚物
VC/OA	Vinylchloride octylacrylate copolymer	氯乙烯-丙烯酸辛酯共聚物
VC/P	Vinylchloride propylene copolymer	氯乙烯-丙烯共聚物
VC/VAC	Vinylchloride vinylacetate copolymer	氯乙烯-乙酸乙烯酯共聚物
VC/VDC	Vinylchloride vinylidene chloride copolymer	氯乙烯-偏氯乙烯共聚物

续表

缩写	英文名称	中文名称
VE	Vinyl ester resin	乙烯基酯树脂
VLDPE	very low density polyethylene	极低密度聚乙烯
VC/E/MA	Vinylchloride ethylene maleic acid copolymer	氯乙烯-乙烯-马来酸共聚物

Appendix 4　Speaking of Common Molecular Formulas, Mathematical Symbols and Greek Alphabet

常用符号	英文名称
$\rightarrow$	give; yield; produce, form, become
$\uparrow$	evolved as a gas; give off a gas
$\downarrow$	is precipitated, gives X precipitate
$\rightleftharpoons$	reacts reversibly
$\xrightarrow{Li, \triangle}$	in the precence of lithium as a catalyst on heating
$CO_3^{2-} + Ca^{2+} = CaCO_3 \downarrow$	a carbonate anion with a balancy of two plus a calcium cation with a balancy of two produce a calcium carbonate precipitate
R'	R prime
R''	R double prime; R second prime
R_1	R sub one
100℃	one hundred degrees ℃
$+$	plus; positive
$-$	minus; negative
$\times$	multiplied by; times
$\div$	divided by
$\pm$	plus or minus
$=$	is equal to; equals
$\equiv$	is identically equal to
$\approx$	is approximately equal to
( )	round brackets; parentheses

续表

常用符号	英文名称
[ ]	square brackets
{ }	braces
$a \gg b$	a is much greater than b
$a \geqslant b$	a is greater than or equal to b
$a \propto b$	a varies directly as b
$\log_n X$	$\log X$ to the base n
$\sqrt[3]{x}$	the cubic root of x
$\sqrt[n]{x}$	the nth root of x
x^2	x square; x squared; the square of x
x^n	x to n factors; the nth power of x; x to the power n
x^{-6}	x to the minus sixth power
$|x|$	the absolute value of x
$\bar{x}$	the mean value of x
Σ	the sum of the terms indicated; summation of
Δx or δx	the increment of x
$\mathrm{d}x$	differential x
$\mathrm{d}y/\mathrm{d}x$	the first derivative of y with respect to x
$\int$	integral
∞	infinity
1/2	a half; one half
2/3	two thirds
6/121	six over a hundred and twenty-one
$6\frac{3}{4}$	six and three over fourths; six and three quarters
0.01	zero point zero one; nought point nought one
6%	6 percent
3‰	3 per mille
2∶3	the ratio of two to three
$r = xd$	r equals x multiplied by d
$5 \times 2 = 10$	five times two equals ten

续表

常用符号	英文名称
$x^3/6=y^2$	x raised to the third power divided by six equals y squared
$(a+b-c\times d)/e=f$	a plus b minus c multiplied by d, all divided by e equals f
$y=(W_t-W)/xy$	y equals W sub t minus W over x

希腊字母		读音	希腊字母		读音
α	A	Alpha	ν	N	Nu
β	B	Beta	ξ	Ξ	Xi
γ	Γ	Gamma	o	O	Omicron
δ	Δ	Delta	π	Π	Pi
ε	E	Epsilon	ρ	P	Rho
ζ	Z	Zeta	σ	Σ	Sigma
η	H	Eta	τ	T	Tau
θ	Θ	Theta	υ	Υ	upsilon
ι	I	Iota	φ	Φ	Phi
κ	K	Kappa	χ	X	Chi
λ	Λ	Lambda	ψ	Ψ	Psi
μ	M	Mu	ω	Ω	Omega

Appendix 5 Common Glassware Names

英文名称	中文名称	英文名称	中文名称
adapter	接液管	boiling flask 3-neck	三口烧瓶
air condenser	空气冷凝管	burette clamp	滴定管夹
beaker	烧杯	burettes tand	滴定管架
beaker tongs	烧杯钳(夹)	Busher funnel	布氏漏斗
boiling flask	烧瓶(夹)	centrifuge tube	离心管

英文名称	中文名称	英文名称	中文名称
claisen distilling head (adapter)	减压蒸馏头	pinch clamp	弹簧节流夹
clamp holder	持夹器	pipette	吸液管
condense-allihn type	球形冷凝管	plastic squeeze bottle	塑料洗瓶
condenser-west tube	直形冷凝管	reducing bush	大变小转换接头
crucible tongs	坩埚钳	rotary evaporator	旋转蒸发器
crucible with cover	带盖坩埚	rubber pipette bulb	洗耳球
distilling head (adapter)	蒸馏头	screw clamp	螺旋夹
distilling tube	蒸馏管	separatory funnel	分液漏斗
Erlenmeyer flask	锥形瓶	spatula	压舌刀;调药刀
evaporating dish (porcelain)	瓷蒸发皿	stemless funnel	无颈漏斗
filter flask (suction flask)	抽滤瓶	stirring rod	搅拌棒
Florence flask	平底烧瓶	stopper	塞子
fractionating column	分馏柱	test tube holder	试管夹
funnel	漏斗	test tube	试管
Geiser burette (stopcock)	酸式滴定管	Thiele melting point tube	提勒熔点管
graduated cylinder	量筒	transfer pipette	移液管
Hirsch funnel	赫氏漏斗	tripod	三脚架
long-stem funnel	长颈漏斗	UV quartz cuvette	紫外石英比色皿
medicine dropper	滴管	vacuum pump	真空泵
Mohr burette for use with pinchcock	碱式滴定管	volumetric flask	容量瓶
Mohr measuring pipette	量液管	watch glass	表面皿
mortar	研钵	water aspirator pump	水抽气泵
pestle	研杵	wide-mouth bottle	广口瓶

Reference

[1] Sorrell T N. Organic Chemistry [M]. 2nd ed. California: University Science Books, 1999.
[2] Carey F A. Organic Chemistry [M]. 3rd ed. Columbus: The McGraw-Hill Companies, 1996.
[3] Allcock H R, Lampe F W, Mark J E. Contemporary Polymer Chemistry [M]. 3rd ed. Beijing: Science Press and Person Education North Asia Limited, 2003.
[4] Seymour R B, Carraher C E J. Polymer Chemistry-an Introduction [M]. 2nd ed. New York: Marcel Dekker. Inc., 1988.
[5] Rosen M J. Surfactants and Interfacial Phenomena [M]. 3rd ed. New York: John Wiley & Sons. Inc., 2004
[6] Zettlemoyer A C. Colloid and Surface Chemistry [J]. Annual Review of Physical Chemistry, 1958 (9): 439-468.
[7] AATCC Publications Committee. The Application of Vat Dyes. AATCC Monograph No. 2 [M]. American Association of Textile Chemists and Colorists, Research Triangle Park, NC, 1953.
[8] BEECH W F. Fibre-Reactive Dyes [M]. New York: SAF International Inc., 1970.
[9] Duffield P A, Lewis D M. Wool Dyeing [M]. Bradford: Society of Dyers and Colourists, 1992.
[10] Corban B P. Textile Fibres to Fabrics [M]. 6th ed. New York: Gtegg Dvision/McGraw-Hill Book Company, 1983.
[11] Mariory J L. Introductory Textile Science [M]. New York: CBS College Publishing, 1981.
[12] Corban B P. Textile Fibers to Fabrics [M]. 6th ed. New York. Gregg Dvision/McGrew-Hill Book Company, 1983.
[13] Wingate I B, Mohler J F. Textile Fabrics and their Selection [M]. New Jersey, Prentice-Hall INC, 1984.
[14] Billmeyer, F. W. Jr., M. Saltzman. Principle of Color Technology [M]. 2nd ed., John Wiley and Sons, New York, NY., 1981.
[15] Diano Corporarion. Color Technology and Its Applications in Industry [M]. Diano Corporation, Foxboro, MA,1970.
[16] Kuehni, R.. Computer Colorant Formulation [M]. D. C. Heath and Company, Lexington, MA, 1975.
[17] Troman E R. Dyeing and Chemical Technology of Textile Fibres [M]. Lodon: Charles Griffin & Company Ltd,1984.
[18] Peter R H. Textile Chemistry [M]. 3rd ed. New York: Elsevier Scientific Publishing Company, 1975.
[19] Steinhardt I, Harris M. Combination of Wool Protein with Acid and Base [J]. J. Res, Nat. Bur. Standard, 1940, 24: 335-367.
[20] Trotman S R. Starch, Dextrin, Glue, Gelatin, Casein and Vegetable Gums, Textile Analysis [M]. London: Griffin, 1932.
[21] Johnson J C. Enzymatic Conversion of Starches, Industrial Enzymes [M]. N. J.: Noyes Data Corp., 1977.
[22] Peters R H. Impurities in Fibres, Purification of Fibres (Vol. II) Textile Chemistry [M]. New York: Elsevier Scientific Publishing Co., 1967.
[23] Trotman E R. Detergents and Scouring, Dyeing and Chemical Technology of Textile Fibre [M]. London: Arnold Publishers, 1990.
[24] Tomasino C. Chemistry and Technology of Fabric Preparation and Finishing [M]. Raleigh: North Carolina State Univ., 1992.
[25] Trotman S R. Bleach Agents, Textile Analysis [M]. London: Griffin, 1932.
[26] Marsh J T. An Introduction To Textile Bleaching [M]. London: Chapman & Hall London, 1948.
[27] Lewin M, Sello S B. Handbook of Fiber Science and Technology: Vol. (I), Chemical Processing of Fibers Processing of fiber and Fabrics Fundamentals and Preparation (part A) [M]. New York: Marcel Dekker Inc., 1996.
[28] Tomasino C. Chemistry and Technology of Fabric Preparation and Finishing [M]. Raleigh: North Carolina State Univ., 1992.
[29] Marsh J T. Mercerizing [M]. London: Chapman & Hall London, 1948.
[30] Hearle J W. The Setting of Fibers and Fabrics [M]. Watford: Merrow Publishing Co., 1971.

[31] Troman E R. Dyeing and Chemical Technology of Textile Fibres [M]. Lodon: Charles Griffin & Company Ltd, 1984.

[32] Peter R H. Textile Chemistry [M]. 3rd ed. New York: Elsevier Scientific Publishing Company, 1975.

[33] Steinhardt I, Harris M. Combination of Wool Protein with Acid and Base [J]. J. Res, Nat. Bur. Standard, 1940, 24: 335-367.

[34] Perkins W S. Textile Coloration and Finishing [M]. North Carolina: Carolina Academic Press, 1996.

[35] Vigo T. Textile Processing and Properties [M]. US Department of Agricultural Research Service, Southern Regional Center, 1100 R. E. Lee Boulevard, New Orleans, LA, USA, 1994.

[36] White M. Developments in Jet Dyeing [J]. Rev. Prog. Coloration, 1999, (29): 94.

[37] Hickman W S. Steam and Steamer [J]. Rev. Prog. Coloration, 1998, (28): 39.

[38] Miles L W C. Textile Printing [M]. 2nd ed. Bradford Yorkshire: Society of Dyers and Colourists, 1994.

[39] Perkws W S. Textile Coloration and Finishing [M]. Durham: Carolina Academic Press, 1996.

[40] Collier B J, TortorA P G. Understanding Textiles [M]. 6th ed. Hertfordshire: Prentice-Hall International Limited, 2001.

[41] 王鸿儒. 制革专业实用英语 [M]. 北京: 中国轻工业出版社, 2005.

[42] Speciality English for Students Majoring in Leather Chemistry and Technology [M]. Xi'an Northwest Institute of Light Industry, 1989.

[43] 曹邦威, 张东成. 制浆造纸专业英语 [M]. 北京: 中国轻工业出版社, 2006.

[44] Smook G A. Handbook for Pulp & Paper Technologists [M]. Vancouver: Angus Wilde Publications Inc, 1992.